Sustainable Breeding in CATTLE & BUFFALO

Sustainable Breeding in CATTLE & BUFFALO

VIKAS VOHRA
M.V.Sc, PhD

Assistant Professor
Division of Animal Genetics and Breeding, F. V. Sc & A.H., Sher-E- Kashmir University of Agricultural Sciences & Technology -Jammu (J&K) - India

A. K. CHAKRAVARTY
M.V.Sc, PhD

Principal Scientist, Dairy Cattle Breeding Division, National Dairy Research Institute, Karnal (Haryana) - India

2011

SSPH

SATISH SERIAL PUBLISHING HOUSE
403, Express Tower, Commercial Complex, Azadpur, Delhi-110033 (India)
Phone : 011-27672852, Fax : 91-11-27672046
E-mail :info@satishserial.com, hkjain1975@yahoo.com
Website : www.satishserial.com

Published by :

SATISH SERIAL PUBLISHING HOUSE
403, Express Tower, Commercial Complex, Azadpur, Delhi-110033 (INDIA)
Phone : 011-27672852 Fax : 91-11-27672046
E-mail : info@satishserial.com, hkjain1975@yahoo.com

ISBN 81-89304-98-4

Typeset at: **Harminder** *for* **Laxmi Art Creations** # 9811482328
Printed at: Salasar Imaging Systems, Delhi-35

Dedicated to

Animals Owners & Breeders

"If I can see far, It is only because I can stand on the shoulders of Giants"
—Sir Isaac Newton

PREFACE

In the recent past landless and small farmers in India committed suicide as they were unable to sustain their livelihood from farming during the famine conditions in some parts of the country, where monsoon was erratic. According to an estimate in the same region, none of those farmers committed suicide who were keeping 2-3 dairy animals. This clearly emphasizes the importance of dairying in our country. Therefore, it is evident that scientific breeding, feeding and management can save many lives by making the dairy sector sustainable and profitable. Sustainable breeding of cattle and buffalo can play a larger role in overall improvement in the production performance for the years to come.

The major thrust of this book is a challenge not only to re-look into the existing dairy animal improvement programmes being followed in the country with various supportive technologies and available resources; but also focused to identify the critical gaps affecting the productivity in cattle and buffalo; and how to mitigate those gaps through intervening improved and sustainable emerging technologies / package of practices so that the productivity of farm animals can be further enhanced and made economically profitable and could be sustained both at organized farms and under rural household conditions. The present book is a coordinated effort of both of us and we place on record our sincere thanks to all authors whose work we have included in *Sustainable Breeding in Cattle and Buffalo.*

The goal of this book is to review and organize the literature dealing with breeding, nutritional & managemental strategies and impact assessment with special reference to the sustainable animal production system. Accordingly, the book is divided into three parts. Part one includes the genetic approach to sustainable breeding of cattle and buffalo, part two includes the feeding and management approach to sustainability and part three consists of assessing the impact of sustainable breeding for higher and sustainable production in cattle and buffalo. There are total nineteen chapters covering systematically these three aspects and to comprehend the ideas related to sustainable breeding.

No attempt has been made to give the exhaustive references to published work; to have done this would have required a much longer book, and a considerable sacrifice of clarity. Although nearly all the work presented in the book can be found in scientific literature but that is scattered in a lot of books, journals, reviews and symposia.

This book targets all the Post Graduate and PhD students including Teachers, Trainees and Professionals of Animal Genetics & Breeding in state, central agricultural universities and other institutes of repute. We sincerely hope that they will find this book useful for various courses in the subject. However, despite the several rounds of reading, there may be technical or printing errors, we will appreciate, if these errors or omissions are brought to our notice, so that future editions may become free of errors and book may prove more useful to the readers. To aid the reader, an alphabetical glossary of terms used is presented after the text.

Authors

ACKNOWLEDGEMENT

We extend our respect and gratitude to Dr. B. Mishra, Vice Chancellor, Sher-e-Kashmir University of Agricultural Sciences and Technology, Jammu (J&K) and Dr. A.R. Nazki, Dean F.V.Sc & A.H. for their kind support and guidance.

We wish to express our deep sense of profound respect and gratitude to Dr. A. K. Srivastava, Director and Vice Chancellor, National Dairy Research Institute (N.D.R.I), Karnal - Haryana, for his vision, dynamic leadership and encouragement that has inspired us in writing this manuscript.

It is our pleasure to thank all those who helped, inspired, encouraged and gave useful suggestion throughout writing this book especially Dr V. D. Mudgal, Ex Director, CIRB, Hisar (Haryana), Dr . V. S. Raina, and Dr P. K. Nagpal, Retd. Principal Scientists, NDRI, Karnal (Haryana), Dr. D. K. Sadana, Principal Scientist, NBAGR, Karnal (Haryana), Dr. A. K. Gupta, Principal Scientists and Dr. I. D. Gupta, Principal Scientists, DCB Division, NDRI, Karnal (Haryana).

We also owe much to all our scientific colleagues in the Dairy Cattle Breeding Division of NDRI, Karnal and Division of AGB, Faculty of Veterinary and Animal Sciences, SUKAST-Jammu.

We sincerely thank Madam Ritu Chakravarty, Sr. Scientist, Dairy Extension Division and Madam Alka Chopra, DCB Division, NDRI, Karnal (Haryana), for meticulous proof reading of the book.

Our mere words will never match the quantum of love, affection and inspiration, which we received from our family members, relatives and friends without their emotional support, especially Virender Bansal and Vishal Bindal, this task would have been truly impossible.

It has been pleasure working with Satish Serial Publishing House, New Delhi.

Above all, we thank Almighty for giving us enough strength for writing this book.

Date: November, 2010 **Authors**

CONTENTS

Introduction

India is a diversified country with a large distinction in agro climatic and economic conditions. Agriculture and its allied activities including animal husbandry and fisheries, continues to be the main source of income in India, with about 58% population of the country dependant directly or indirectly on this sector. Livestock sector has emerged as the fastest growing sector of the Indian agriculture in the last decade. At present milk is the prime contributor to agricultural GDP.

There are well-adapted indigenous cattle and buffalo breeds for milk, dual and draft animal power in the country. In India cattle and buffaloes are reared for the most part under low input system. The bulk of milk is sold under unorganised markets in India and negligible percent of milk is processed into value added products. The landless and marginal farmers of the country are able to maintain their regular expenses through the sale of raw milk from cattle and buffaloes in the local market, which bring liquid cash to meet their routine expenses and help to sustain their livelihood in between the earnings from crops, which are seasonal. The productivity of dairy animals in India is low as compared to the temperate breeds of developed countries. The reason attributed to this is that we are still short in green fodder and concentrates, to an extent that we are not able to feed our animals even their maintenance requirements, which decreases their performance. Other reasons include high calving intervals, low fertility and management is not adequate. Despite these facts, India enjoys the number one position in the world in terms of overall milk production (~106 million tonnes, 2010) because of good germplasm and large number of dairy animals.

Breeding large ruminants for meat purpose is not practised in India. Cattle rearing for beef production is not allowed in India (barring few states), as cows are regarded sacred and have religious sentiments in the country. Buffalo meat is produced commercially in the country, but so far, this animal has not breed especially for meat purpose. Draft power for agricultural operations and transport is mainly supplied by cattle and buffalo bulls. The dung is used as organic manure for maintaining the soil fertility on long term basis and for providing cooking fuel. Livestock rearing has been an important source of livelihood for landless labourers and small farmers. For tribal and nomads, animal rearing / keeping is the only source of earning. Therefore, livestock sector has a potential, and will play the larger role in the upliftment of rural poor.

While developing sustainable breeding plan for genetic improvement of dairy animals, few things should be kept in mind. Firstly, location specific parameters, (especially inputs, type of breeds to be used and socio economic conditions of farmers) need to be given priority. Secondly, the indigenous well defined breeds of cattle and buffalo need to be improved through *selective breeding* only. The magnitude to which non descript population are to be graded up / crossbred should be decided first and the production of quality dairy bulls in the required numbers should be worked out in detail, so that appropriate programmes for their multiplication and improvement are drawn up because so far, the existing set up of small independent farms of indigenous breeds has not given desired results. Cross breeding of local and non-descript cows with Holstein and Jersey bulls is an alternative and is recommended under breeding policy of cattle in the country. In order to meet the growing demand of milk, a number of synthetic cattle breeds namely Karan Fries, Karan Swiss, Frieswal, Sunandini, Phuletriveni, Jersind etc., have been developed in the country. These synthetic breeds have shown high performance under optimum inputs and are well adapted to local climatic conditions. For sustained breeding and subsequently for high milk production from these synthetic breeds, it is essential that these are multiplied and produced in large numbers under field conditions. Wrong implementation of the cross breeding policy in the past, has resulted in mixture of exotic germplasm into well defined indigenous cattle breeds of the country. On the other hand, grading up (not cross breeding) is strongly recommended for improving non-descript buffaloes. Therefore, breeding programmes for improving indigenous breeds of cattle and buffalo, including crossbreds needs clear understanding. The policy makers should develop a concrete system, which would allow selection within the population to be efficient and more effective. Thirdly, the change of mind set of farmers, from natural mating to artificial insemination is needed. Artificial insemination can have great impact on dairy animal breeding as this will provide additional opportunities for genetic improvement where good sires are not available. Breeding policy should be directed to cover all the breedable dairy animals, through artificial insemination especially in buffalo. Animal breeders and veterinarians need to put their efforts toward the design of breeding programs for artificial insemination through aggressive campaign with farmer's cooperation.

Sustainability

"Sustainable development is the management and conservation of the natural resource base and orientation of the technological and institutional change in such a way as to ensure the attainment and continued satisfaction of human/animals needs for present and future generations" - FAO.

Sustainable development conserves land, water, plant and animal genetic resources. It is environmentally non-degrading, technically appropriate, economically viable and socially acceptable. The consideration of sustainability implies the use of resources very judiciously in such a way that these resources are not exhausted and are left for use for the subsequent generations.

Animals play an essential role in faster recycling of non-food bio-mass and contribute significantly to the economic stability of various farming systems. Therefore, sustainable animal production and development needs urgent attention.

Determinants of Sustainable Animal Production

The development and planning for sustainable livestock breeding and production is the function of market demand for milk and animals products, use and spending channels, pricing potential, interface of land & animal resources, transferable technologies and their impact, sustainability of natural resources and their effect on environment and finally the manpower. The manpower involves the farmers, govt. and private support personnel involved in research, training, extension, market regulation, finance/policy formulation and implementation. The needs of the farmers are a function of tradition, education and day-to-day survival demands influenced by the aspirations of individual, family and the local community. The planning of any breed improvement programme has to take care of local tradition, culture and status of the farming community. The planner must perceive the practical problems of the farmers and should give solutions for these problems. The technologies developed by researchers should enhance the production efficiency of farming community and simultaneously should be cost effective, easy to use, socially acceptable and eco-friendly. In our country, the research and extension services are not sufficient to support livestock development programme. Staff motivation and research-extension-support services links have not given adequate attention by planners, and needs to be addressed.

Following are the determinants of sustainable animal production system

1. Breeding programmes undertaken for livestock improvement
2. By judicious use of available land and feed resources
3. Traditionally available feeds/fodder should be used along with supplementation of diet with minerals, proteins or energy limiting rations.
4. Improvement of feed utilization through the treatment of by-products (with ammonia, urea), diet supplementation with high energy feeds (urea, molasses etc.) coupled with by-pass protein would further enhance the animal productivity.
5. In the long term planning, there is need to apply bio-techniques like development of high biomass feed resources and manipulation of rumen micro flora to improve the digestion of high fibre feeds/fodder in ruminants with emphasis on low methane production system.

6. Role of developmental and planning agencies to undertake and execute the development programmes effectively
7. Motivate the farmers/livestock keepers, who are the end users of technologies developed by various agencies.
8. Multi-purpose farming systems in which integrated livestock, food and tree crop production maximizes photosynthetic capture to produce food, fibre, energy and fuel wood while maintaining the soil fertility and overall sustainability within the system must be evolved for sustainable dairy production in cattle & buffaloes.

The concept of sustainable livestock breeding and production throws many challenges to livestock keepers, research and development agencies, financial institutions, national planners and finally to the consumer of animal products. Emphasis is needed to be given to evolve a multi-faceted approach which can make livestock production in our country more sustainable enterprise. Keeping in mind, the multifaceted approach to sustainability, this book provides an insight of sustainable breeding in cattle and buffaloes. The book is divided into three parts, Part I is covered in first 12 chapters which deal with the genetic and breeding interventions for sustainable milk production from cattle and buffaloes. Part II comprises of chapters 13 to 15 and deals with the nutrition, methane emission and management interventions for sustainable milk production. Part III of the book deals with how to assess the dimensions of sustainability and this topic is covered from chapter 16 to 19.

Sustainability is the notion and the need of the day in Indian agriculture and animal husbandry. India has majority of the world's cattle and buffalo, also the number one milk producer in the world. Consequently, sustainable breeding of existing cattle and buffalo breeds, using progeny tested bulls at organised herds and field conditions, conserving our native breeds by making them more sustainable and productive, critical evaluation and documentation of our existing livestock are few points kept in mind while writing this book. If due importance is given to these aspects and an effort is made to mitigate the gaps and shortcoming. I am sure that cattle and buffalo of India can lead her to new heights in the dairy sector.

❑❑❑

Part-1
Genetic Interventions in Sustainable Breeding

Chapter 1

Genetic Resources of Indian Dairy Cattle and Buffalo

Origin and Distribution

To pin point the exact period for origin of cattle and buffalo is very difficult since there are different references. Based these on one can conclude that cattle were first domesticated along with sheep after the domestication of dog and goat. The buffalo were originated in India about 4000 years ago. In Indian history the oldest reference of cow is documented in Rig Veda as 'Kam dhenu'. In hindu mythology the buffalo bull has been referred to as vehicle of 'Yamraj' the God of death. The Indian cattle and buffalo breeds are known for their hardiness, their ability to digest coarse fibres, endurance for heat tolerance, resistant to tropical diseases and adaptability to the tropical harsh climatic conditions. Indian cattle are excellent draft purpose breeds. On the other hand, India possesses the best germplasm of dairy buffaloes in the world.

The Cattle is a subfamily, known as bovinae, of the family bovidae of horned animals, having 9 subspecies of cattle under genus *Bos*. These are Zebu cattle, humpless cattle, buffaloes, bison, mithun, yak, banteng, gaur and kouprey. The domestic cattle descended from the aurochs and divided into two sub groups viz. zebu cattle of Asia and Africa and the exotic cattle or humpless cattle predominantly of Europe. The humpless cattle are called *Bos Taurus* and are primarily the milch breeds. The zebu cattle are found in Indian subcontinent belong to draft and dual purpose breeds. About 80% of the total cattle population in India is non-descript or local.

Classification of Cattle Breeds

India has about 130 breeds of domestic animals, out of which 30 breeds are of cattle (milch = 4, dual purpose = 8 and draft type = 18). The major milk breeds are Sahiwal, Gir, Red Sindhi and Tharparkar. The dual purpose breeds of cattle include Hariana, Deoni, Ongole, Rathi, Dangi, Mewati, Kankrej and Krishna Valley, remaining is draft purpose breeds. The cattle of India vary in size and body conformation which is attributed to their origin, adaptation and to different agro climatic conditions and also to the use for which they have been put to. According to National Bureau of Animal Genetic Resources, Karnal, there are Six (6) breeds of cattle which are considered as at risk namely, Punganur, Vechur, Krishna Valley, Red Sindhi, Sahiwal and Amritmahal. The number of Punganur, Vechur, Siri, Ponwar, Kenkatha, Nagori, Bachaur, Mewati and Kherigarh cattle is too low that these nine (9) breeds are considered as endangered.

Classification of Buffalo Breeds

India possesses many important breeds of buffaloes. Indian wild buffalo, *Bubalus arnee* receives its specific name from the Hindi 'arni'. Whereas the domestic buffalo was given the specific name, bubalis by Linnaeus. According to Bohlkon's terminology, it should be called Bubalus arni forma bubalis. The domestic buffalo in different countries is known by different names such as "bhains" in India, "aljamoos" in Arab countries, karabue" or "kwai" in Thailand, "carabao" in Phillipines and "karbo in Malaysia. At present, there are about 3213 breeds of livestock enlisted in the dictionary of livestock breeds of which 74 buffalo breeds are available in the world. Most of the buffalo breeds are used mainly for milk production purpose. However, buffaloes also play an important role in the production of meat and draught power. According to I.C.A.R there are fifteen (15) breeds of buffalo out of which seven well defined milch breeds of buffalo's viz. Murrah, Bhadawari, Mehsana, Surti, Jaffarabadi,

NilliRavi and Nagpuri which are contributing significantly in the national milk production. Out of these, there are two buffalo breeds which are at risk namely, Toda and Bhadawari, according to National Bureau of Animal Genetic Resources.

The buffaloes in India have been mainly classified into six distinct groups as

(i) *Murrah group* comprises Murrah and Nili-Ravi which have the home tract in Haryana and Punjab, respectively

(ii) *Gujarat group* comprising Jaffarabadi, Surti and Mehsana, has origin in Gujarat

(iii) *Uttar Pradesh group* has Bhadawari and Terai

(iv) *Central India group* includes Nagpuri breed in Maharashtra, Panndharpuri breed in Maharashtra, Andhra Pradesh and Karnataka, Kalahandi breed in Andhra Pradesh and Orissa and Sambalpuri breed in Orissa and Madhya Pradesh. Besides, Manda and Jerangi breeds also belong to this group.

(v) *South India group* comprises Toda and South Kanara.

(vi) *North-Eastern group* includes Swamp buffaloes mainly available in Assam.

Cattle and Buffalo Population

According to the 17th Indian Livestock Census (2003), India possesses world's largest bovine and bubaline population. India possesses 185.18 million cattle's which is about 14 percent of the world cattle population. Out of the 185.18 million cattle's in India about 24.68 million are crossbred amounting to 13.33 percent of total cattle population. Annual growth rate in total cattle population showed negative trend with -1.18 % from 1997 to 2003. Total number of indigenous and crossbred cattle in 2003 was 160.49 and 24.68 million respectively (Table 1). The annual growth rate of indigenous and cross bred cattle was -1.78 and 5.73% respectively during 1997-2003 (Figure 1). However, a decline of around 19.4 million and 18.29 millions was observed in total cattle population and indigenous cattle respectively. On the contrary, a phenomenal growth of 5.73% (1997-2003) was observed in cross bred cattle.

Buffalo's population is 96.90 millions in India which account for 56.8% of world's bubaline population. Annual growth rate in buffaloes were observed at 1.37% from 1992 to 2003 and at 1.43% during the period from 1997 to 2003. Total number of buffaloes was increased from 89.91 millions in 1997 to 96.90 millions in 2003. The Zone-wise total population a trend shows that buffalo population has increased in all parts of the country. The trend of breedable buffalo population including heifers (1951 to 2000) shows an overall increase of 95.52%.

Thus, it is imperative to survey the breed-wise population of cattle and buffalo at a shorter period of time to assess the dynamics of breed-wise cattle and buffalo population.

Table 1: Cattle Population in India (in millions)

	1992	1997	2003	% increase per annum from 1997 to 2003
Cattle	204.58	198.88	185.18	-1.18
Adult female cattle	64.36	64.43	64.51	0.02
Crossbred	15.21	20.09	24.68	5.73
Indigenous cattle	189.37	178.78	160.49	-1.78

Table 2: Buffalo Population in India (in millions)

	1992	2003	% increase per annum from 1992 to 2003
India	84.20 (54.99)	96.90 (56.80)	1.37
Asia	148.35 (96.89)	165.45 (97.10)	1.04
Rest of the world	4.757 (3.11)	5.01 (2.90)	0.48
Total World	153.107	170.458	1.03

Figures in Parenthesis indicates percentage of World Population

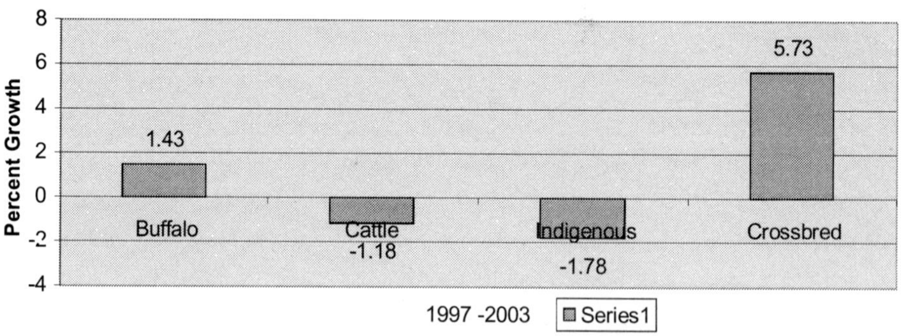

Figure 1: Percent Annual Growth Rate in Dairy Animals

Trends in Milk Production

Milk production is predominantly the domain of landless labourers and small/marginal farmers in our country, who generally keep 1-2 milch animals under mixed farming system. These small holders have about 70% of the milch animals. Majority of the animals are reared on feed stuffs not consumed by human population (wheat and paddy straw, maize, sorghum and millet stovers and other crop by-products and waste material). India has attained number one position in milk production after surpassing USA during the year 1998 (74 versus 71 million tones). There has been laudable increase in milk production in India during the post-independence era. It has increased more than six folds from 17 million tones during 1951 to ~106 million tones during 2009-10. The increased production of milk has improved the per capita availability to approximately 250gms/day which is marginally short of recommended 280gms/day per person by I.C.M.R (1989). It is further projected that by the year 2020 the milk production of India would be 168 million tones. The contribution of our country to global milk production would increase substantially. On considering trend it has been found that annual growth rate of milk production in India is positive with 5.5% at present and nearly with 4 percent since, last ten years as compared to the world annual growth rate of 1.5 percent. According to the estimate of Central Statistical Organization the contribution of livestock and fisheries sector to the agriculture sector is 31.5% of the value of the output. This has now become comparable to the livestock contribution in the world agriculture sector which is about 40%. The contribution of these sectors to the total GDP was 7.35% during 2001-02. The contribution of milk alone in GDP is higher than paddy, wheat and sugarcane.

According to the Department of Animal Husbandry Dairying and Fisheries, Ministry of Agriculture (2006) buffaloes are producing 53.64% of total milk in our country (Figure 2). Cows produce 40.96% of milk, out of which 21.03% is produced by indigenous cattle and 19.92% by crossbred cows. The 5.40% of total milk is produced by the goats. Crossbred cattle produces 19.34 million tones of milk which is 48.65 percent of the total cow milk produced in the country. However, the productivity of milk is much lower as compared to the developed countries of the world. According to an estimate based on comparison of milk production of 44 top ranking countries of the world during 2000, the productivity/animal/lactation in India was 917 kg with its 35th ranking out of 44 countries. Therefore, there is a continuous need to improve the productivity as well the quality of milk in order to be competitive in global scenario of dairy farming. Growth in milk production by crossbred cows is phenomenal (94%) as compared to about 32% from indigenous cows, which indicates that we cannot neglect to maintain crossbred population in certain potential pockets of the country. This also indicated that there is further scope for improving the milk production potential of cattle through systematic implementation of our crossbreeding programmes in the country. The Zone wise milk production shows that only 14 states of India produce about 94.30% of the total bovine milk while remaining 5.70% of bovine milk is produced by 12 other states and 5 Union Territories leaving a wide scope of milk production through systematic development of sustainable cattle and buffalo breeding policies and their implementation through various projects and programmes for increasing the productivity and other performances.

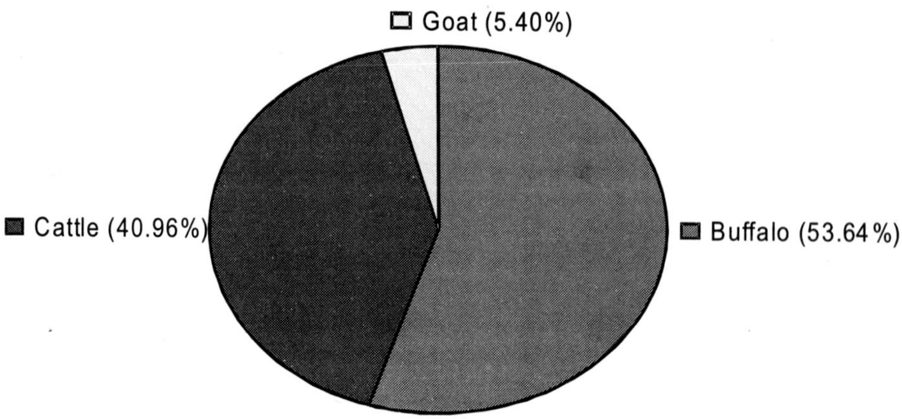

Figure 2: Percent Share of Dairy Animal in Total Milk Production

Major Cattle Breeds of India

Sahiwal

It is the most popular breed of the subcontinent and is Native of Pakistan and its breeding tract is Montogomery district which is now named as Sahiwal. The animals are found fairly in India also. It is comparatively a heavy breed with symmetrical body and loose skin (lola) (Figure 3). The distinguishing feature between sahiwal and red sindhi is the muzzle, red sindhi has dark colour muzzle where as sahiwal has lighter colour muzzle.

Gir

It is a milch breed of India with a native of Gujrat but is also found in Maharashtra and adjacent Rajasthan. It's features are protruding broad and long forehead, and pendulous front wards turned ears (Figure 3).

Red Sindhi

Milch breed of India with its home tract in Karachi and Hyderabad districts of Pakistan, however, a number of herds of this breed are found in certain organised herds / institutions in India. Red Sindhi are small in size and are very good milkers. They have compact frame of body with round dropping quarters (Figure 3). The colour is red with different shades. Stripes of white colours are sometimes seen on dewlap and occasionally on the forehead.

Tharparkar

The origin of this breed is from the district of that name in the province of sind (now in Pakistan). The animals are also known as 'thari'. This breed is also found in the adjoining tracts of Rajasthan, especially around Jodhpur and Jaisalmer. This is a medium sized compact breed (Figure 3) with good milk production. The males are used for draft purpose.

Hariana

The breeding tract of this dual purpose breed is Haryana. The centre of origin is around Rohtak, Hissar and Gurgaon districts. It is widely spread along the indo-gangetic plains and Punjab, and also in some parts of Madhya Pradesh (Figure 3).

Ongole

The native tract of this breed is the Guntur district of Andhra Pradesh. It is a dual purpose breed (Figure 3). This is essentially a large muscular breed suitable for heavy draft work.

Deoni

This is an admixture of Gir, Dangi and local animals of western Andhra Pradesh. The bullocks are suitable for heavy cultivation. It is a dual purpose breed (Figure 3).

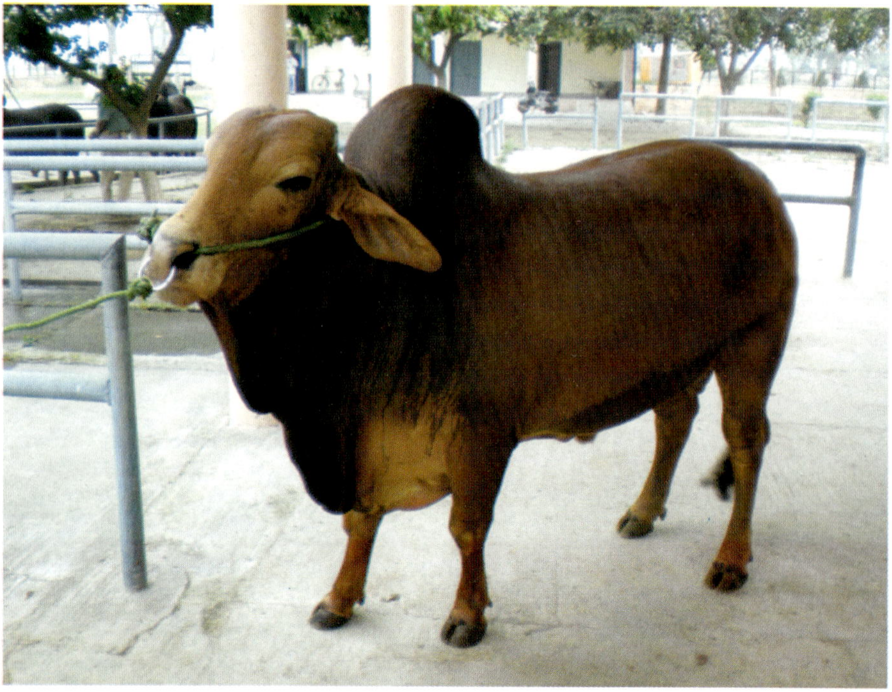

Figure 3 : a) Sahiwal Bull (Courtsey NBAGR, Karnal)

Figure 3 : b) Sahiwal Cow

Figure 3 : c) Gir Bull (Courtsey NBAGR, Karnal)

Figure 3 : d) Gir Cow

Figure 3 : e) Red Sindhi Bull (Courtsey NBAGR, Karnal)

Figure 3 : f) Ongole Bull (Courtsey NBAGR, Karnal)

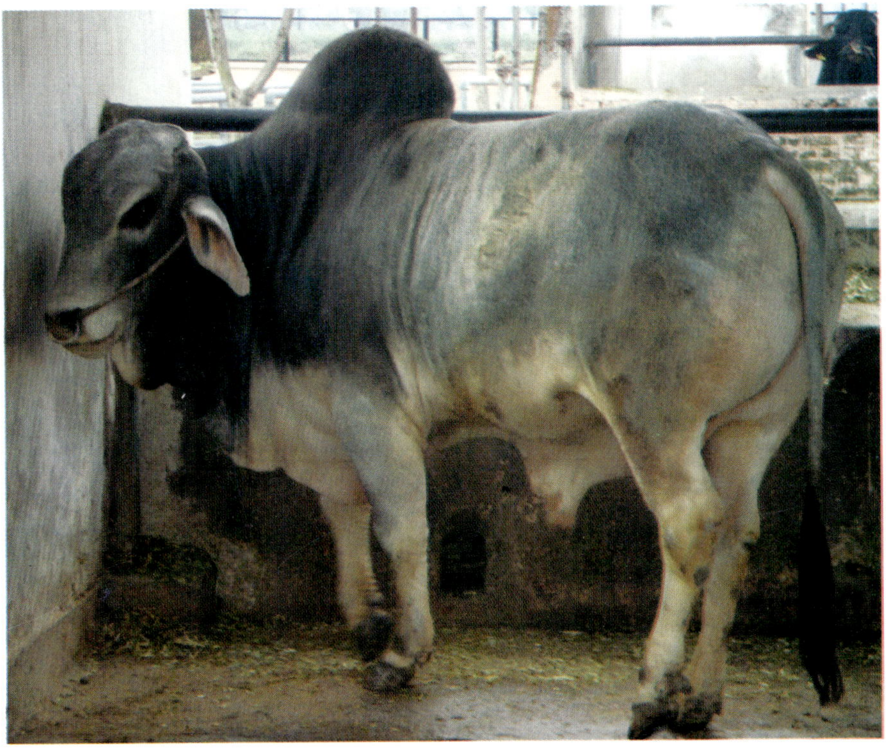

Figure 3 : g) Tharparkar Bull

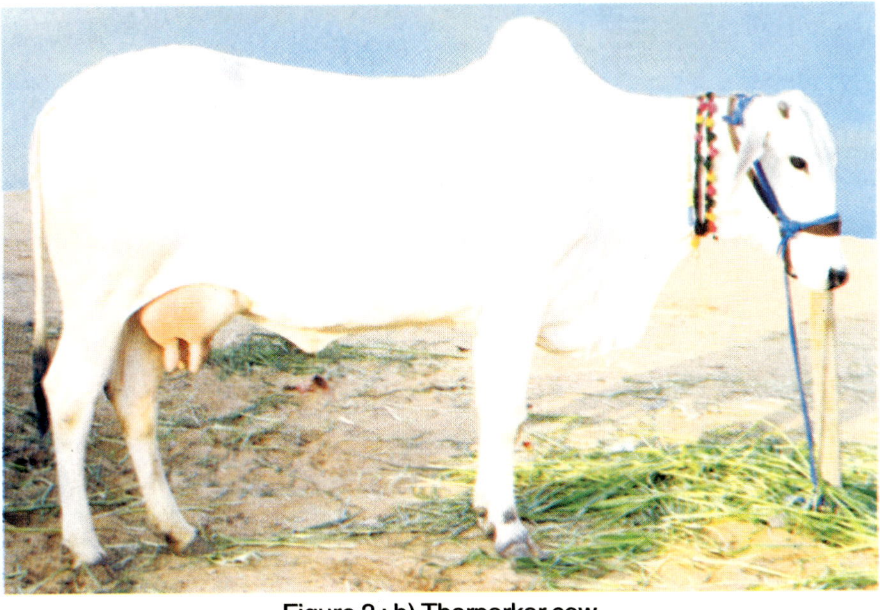

Figure 3 : h) Tharparkar cow

Figure 3 : i) Hariana Cow (Courtsey NBAGR, Karnal)

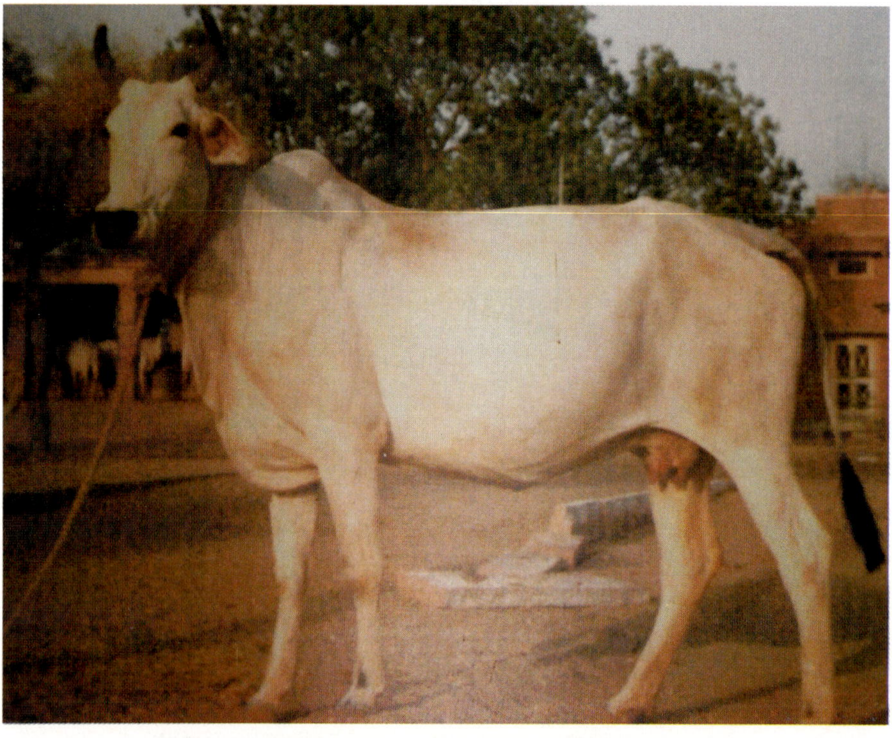

Figure 3 : j) Nagori Cow (Courtsey NBAGR, Karnal)

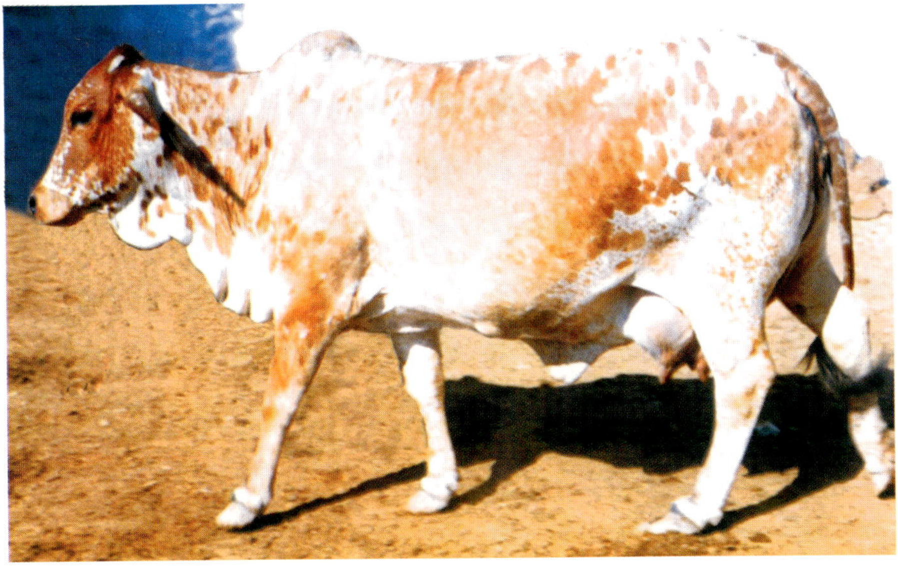

Figure 3 : k) Rathi Cow

Figure 3 : l) Deoni Bull (Courtsey NBAGR, Karnal)

Figure 3 : m) Kankrej Cow (Courtsey NBAGR, Karnal)

Figure 3 : n) Amritmahal Bull (Courtsey NBAGR, Karnal)

Figure 3 : o) Panwar Cows (Courtsey NBAGR, Karnal)

Figure 3 : p) Tho Tho Cow

q) Malarad Gidda (Courtsey NBAGR, Karnal)

Figure 3: Cattle Breeds of India

Major Buffalo Breeds of India

Murrah

Murrah is the world's best riverine buffalo germplasm available in India with the breeding tract in Rohtak, Hissar and Jind district of Haryana, and Nabha and Patiala district of Punjab. The breed characteristics includes massive jet black body, neck and head comparatively long, horns short and tightly curved, udder well developed, hips broad, and fore and hind quarters dropping (Figure 4). The tail is long reaching upto the fetlock. Sometime white markings are found in tail and extremities.

Nili Ravi

The breed is found in Sutlej valley in Ferozpur district of Punjab in India and in the Sahiwal district of Pakistan. The head is elongate, with the peculiar wall eyes; the body colour is black with white markings on forehead, face, muzzle and legs (Figure 4).

Figure 4 : a) Murrah (Courtesy: NBAGR, Karnal)

Figure 4 : b) Nilli Ravi (Courtesy: NBAGR, Karnal)

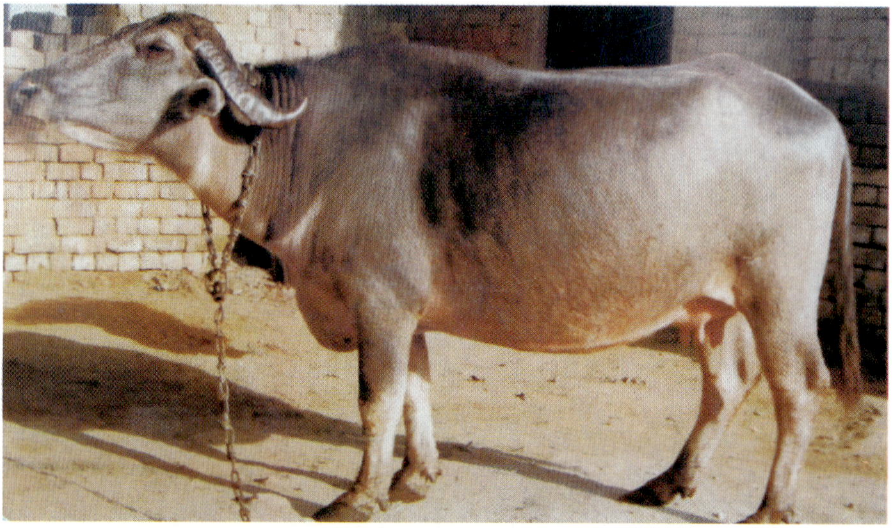

Figure 4 : c) Bhadawari (Courtesy: NBAGR, Karnal)

Figure 4 : d) Jafranabadi (Courtesy: NBAGR, Karnal)

Figure 4 : e) Toda (Courtesy: NBAGR, Karnal)

Figure 4 : f) Surti (Courtesy: NBAGR, Karnal)

Figure 4 : g) Mehsana (Courtesy: NBAGR, Karnal)

Figure 4 : h) Pandharpuri (Courtesy: NBAGR, Karnal)

Figure 4 : i) Nagpuri (Courtesy: NBAGR, Karnal)

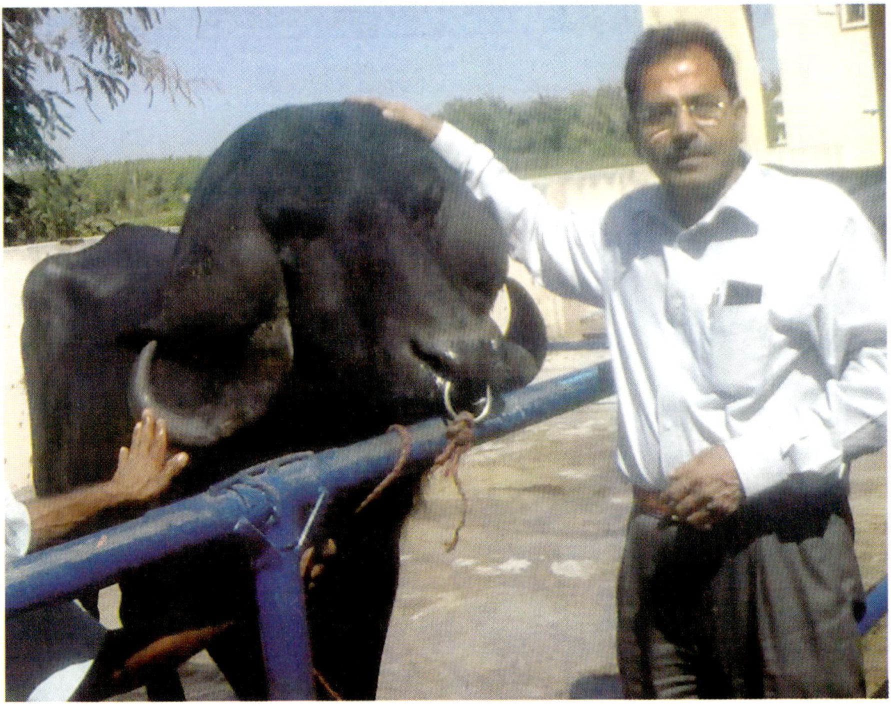

Figure 4 : j) Jafarabadi Bull

k) Murrah Bull
Figure 4: Buffalo Breeds of India

Surti

The breeding tract of this breed is Kaira and Baroda districts of Gujrat. Body is well shaped and medium sized; the barrel is vedge shaped, head is long with prominent eyes, the horns are sickle shaped, moderately long and flat. The colour is black or brown (Figure 4). The pecularity of the breed is two white collars, one round the jaw and the other at the briskit region.

Bhadawari

The breed is found in the bhadawari tehsil in Agra and Etawah district of Uttar Pradesh and Gwalior district of Madhya Pradesh. The characteristic peculiarity of the animal is copper coloured body. Ears are thin (Figure 4). The milk of these animals has highest percentage of fat.

Jafrabadi

The breeding tract of this breed is Kutch, and Jamnagar district of Gujrat, the body is long but not so compact. The dewlap in females is some what loose and the udder is well developed. The horns are heavy, inclined to droop at each side of the neck and then turning up at points, and but less tightly curved (Figure 4). The animals are mostly maintained by traditional breeders called Maldharis, who are nomads.

Nagpuri

The breeding tract of this breed is Nagpur, Akola and Amrawati districts of Maharashtra. This is also called as Elitchpuri or Brari. Horns are long, flat and curved, bending backwards on each side of the back almost to shoulders. The face is long and thin.

Mehsana

Its breeding tract is Mehsana, Sabarkantha and Banaskantha districts of Gujrat. The breed is supposed to have been evolved out of cross breeding between surti and murrah breeds. The body is longer than murrah and limbs are lighter (Figure 4). The head is longer and heavier. Horns are less curved at the end compared to murrah breed but are longer and could be of irregular shape.

The concise summary of growth, production and reproduction characteristics of some of the important breeds of cattle (Table 3), crossbreds (Table 4) and buffalo (Table 5) are presented in the tabular form below.

Draft Animal Power

In India, livestock farming is the backbone of the agriculture and cattle were always considered as the primary source of working bullocks for farm operations. Best draft breeds of Asian Continent namely Nagore, Khillari, Amritmahal, Hallikar and Krishna Valley are available in our country besides dual purpose breeds like Tharparkar, Hariana and Ongole. Mechanization of the farm operations has attributed to the decrease in the draft animal power from 72 to 12% between 1961 and 1995 and further lower during 1995-2005. The working male population among cattle and buffaloes decreased by 14 millions during 1977–87 and the trend continued even in nineties and during early two thousand. The fully grown zebu bullock can provide 0.5 HP and one bullock pair can cultivate about 0.33 ha land in a six-hour working/day. About 8.9 million pairs or 17.8 million bullocks are required to cultivate 176.7 million ha of cultivable land available in India over a period of 60 days (30 days each for rabi and kharif season). Considering a replacement rate of 33% per year about 6 million males are required for replacement, which could be raised from 18 million breedable females. It may be noted that considering draught animal power saves six million tones of petroleum per year, valued at present more than Rs. 24000 crores.

Table 3: Growth, Production & Reproductive Characteristics of Some of the Important Cattle Breeds

Breeds	Birth wt. (Kg)	Wt. at First Calving (Kg)	Age at First Calving (months)	Service Period (days)	Calving Interval (days)	Av. 305 days milk yield (Kg)
Sahiwal	19 - 21 (20.6)	287 - 380 (343)	35 - 45 (41)	136 -189 (177)	418 - 473 (445)	2100
Gir	21 - 23 (21.8)	319 - 327 (324)	44 - 61 (53)	135 - 259 (176)	456 - 541 (511)	1500
Tharparkar	21 - 22 (21.4)	292 - 376 (324)	39 - 53 (43)	130 - 183 (160)	418 - 474 (447)	1700
Red Sindhi	18 - 21 (18.9)	340 - 348 (344)	39 - 49 (44)	152 - 158 (154)	436 - 562 (496)	1600
Hariana	21 - 23 (21.7)	287 - 370 (324)	41- 49 (51)	136 - 303 (214)	434 - 523 (497)	900
Ongole	NK	NK	36 - 42 (39)	210 - 241 (226)	529 - 637 (553)	700
Deoni	NK	NK	47 - 55 (51)	173 - 192 (184)	456 - 472 (466)	900

Figures in the parenthesis shows average values
NK : Not Known

Table 4: Production and Reproductive Characteristics of Some of the Important Crossbred Cattle

Synthetic Breeds	Developed by	Age at First Calving (months)	Service Period (days)	Dry Period (days)	Calving Interval (days)	Lactation Length (days)	Av. 305 days milk yield (Kg)
Karan Fries	N.D.R.I (1981)	32.6	129	68	404	337	3903
Karan Swiss	N.D.R.I (1980)	34.6	115	62	399	306	3615
Frieswal	MDF +ICAR (1985)	31.9	146	-	444	341	2971
Sunandini	Indo Swiss project (1983)	30.5	140	-	420	-	2901

Figures in the parenthesis indicates year of development; NDRI: National Dairy Research Institute, Karnal (Haryana); MDF : Military Dairy Farm, Meerut (U.P); ICAR : Indian Council of Agriculture Research, New Delhi

Table 5: Growth, Production and Reproductive Characteristics of Some of the Important Buffalo Breeds

Breeds	Birth wt. (Kg)	Wt. at First Calving (Kg)	Age at First Calving (months)	Service Period (days)	Calving Interval (days)	Wet Average (Kg)	Av. 305 days milk yield (Kg)
Murrah	27 – 41 (34)	446.4	40 – 44 (42)	113	480	7.5	2000
Nili Ravi	27 – 41 (34)	531.1	42 – 46 (44)	203	570	7.5	1800
Surti	24 – 30 (27)	NK	46 – 54 (50)	NK	NK	6.5	1600
Bhadawari	24 – 30 (27)	425	46 – 54 (50)	203–257 (230)	535–555 (545)	5.0	900
Nagpuri	NK	300	36 – 42 (39)	120–140 (130)	425–445 (435)	6.0	1000
Jafarabadi	NK	NK	42	140	507	7.0	2000

Figures in the parenthesis shows average values
NK : Not Known

Summary

Cattle were first domesticated along with sheep after the domestication of dog and goat. In India, there are 30 breeds of cattle (milch = 4, dual purpose = 8 and draft type = 18). Cattle breeds which are at risk include Punganur, Vechur, Krishna Valley, Red Sindhi, Sahiwal and Amritmahal. Endangered cattle breeds are Punganur, Vechur, Siri, Ponwar, Kenkatha, Nagori, Bachaur, Mewati and Kherigarh. Annual growth rate of indigenous and cross bred cattle was -1.78 and 5.73% respectively, during 1997-2003. One cannot neglect to maintain crossbred cattle in certain potential pockets of the country where good feed and fodder is available, as their contribution is significant in national milk economy. The buffalo was originated in India about 4000 years ago. Murrah buffalo is the world's best germplasm available with India and their breeding tract is in Rohtak, Hisar and Jind district of Haryana, and Nabha and Patiala district of Punjab. Seven well defined milch breeds of buffalo's found in India are Murrah, Bhadawari, Mehsana, Surti, Jaffarabadi, NilliRavi and Nagpuri and they contribute significantly in the national milk production. Two buffalo breeds which are at risk, namely Toda and Bhadawari. Total number of buffaloes was increased from 89.91 millions in 1997 to 97.92 in the year 2003.

□□□

Chapter 2

Riverine Buffaloes-Biggest Potential in Sustainable Dairy Production

Buffaloes are considered as the main dairy animal in India. They are also the important source of meat and draft power. The world's buffaloes are classified into two groups The African Buffalo (Syncerus) which are wild and Asian Buffalo (*Bubalus*) which are mostly domesticated. The Asian buffalo consists of three species the 'anoa' of Celebes, the 'tamaro' of Mindori and the 'arni' or Indian wild buffalo. *Bubalus bubalis* is the name of the domesticated Indian buffalo, previously called *B. arnee*, representatives of which are still found in the jungles of Assam (India). There are two general types of the domesticated buffalo, the river buffalo and the swamp buffalo. The river buffalo has 50 chromosomes and the swamp type has 48, the amount of genetic material is the same in both. They interbreed and produce fertile progeny with 49 chromosomes. The African buffalo is named Syncerus with the species *S. caffer*, with a small number of sub-species. The African buffalo has not been domesticated and will not be discussed further. The *B. Bubalis* is widely distributed in Asia, but it has also been introduced to Europe, Near East, China, South America, the former Soviet Union and the Caribbean. Over half (53%) the world's buffalo population is in India which is approximately 90 million animals.

Swamp Buffaloes

It has received this name from its natural habitat which is swamp or marshland. Swamp buffaloes resemble the wild 'arnee' in morphological characteristics. The swamp type is most common in Southeast Asia where it is mostly used as a draft power for rice cultivation in paddy fields. Swamp buffaloes have a very low milk yield (approximately 400 -500 kg / lactation) and are not used as milk producers. Swamp buffalo tend to be indigenous to those parts of Asia which do not have a great culture from drinking milk or eating milk based products. Recently, its potential as a meat producer has been discovered. Crossings between river and swamp buffaloes have been attempted in Thailand, Philippines, Vietnam, and China on a large scale. These crossbred buffaloes are powerful work animals, and produce good quality meat and relatively more milk.

River Buffaloes

It has received this name from its natural habitat which is their preference to wallow in clean water and rivers. River buffalo tend to be found in those countries where milk plays a more important part in the human diet. This ranges from India through Pakistan into the Middle East and in Bulgaria, Hungary, Turkey, Italy and Egypt. They are also found in Brazil and Caucasia. River buffaloes have been selected for milk production to a larger extent and some river breeds. E.g. Murrah, Nili-Ravi and others are solely used for milk production. In northern India buffalo milk is more preferred over the cow milk. River buffaloes vary greatly in size. The single greatest reason for this variation is environmental. In areas where animal feed is scarce then animals tend to be smaller than in those where it is more plentiful. Typically adult females can range anything between 350kg, high in the Himalayas, to 800kg in Bulgaria and Italy. The buffaloes found in India are comparatively heavier where a male weighs about 1100 kg and female 550 kg. As buffalo are capable of surviving on very little they are often fed very little. Latest estimates of the world population of water (domestic) buffaloes are around 150,000,000 animals in 50 countries worldwide.

Buffaloes are perfectly suited to the hot and humid climate with muddy terrain of India; they have a good coat of soft hair like that of cattle at birth and during early calf hood. The hair on the body becomes sparser and almost devoid of hair as the buffaloes grows. The amount of hair coat retained varies considerably, depending upon the breed, season, and housing practices and because of its exposure to water and mud. The skin is covered with a thick epidermis and the basal cells which contains extra melanin particles that give the skin surface its characteristic black colour. Buffaloes have well developed sebaceous glands, with greater secretary activity than cattle. These glands secrete the sebum; a fatty substance emerges on the skin surface and covers with it a lubricant, making it slippery for water and mud. This greasy sebum, along with the thick hornified top layer of skin, prevents water and the solutes in it from being absorbed into the skin. In this way, the animal is protected from the harmful effects of any deleterious chemical compounds in the water. Moreover, the sebum layer melts during hot weather and becomes glossier to reflect many of the heat rays, thus relieving the animal from the excessive external heat load. However, although buffaloes are perfectly suited to their environments, but suffer in the sun, since they exhibit signs of great distress when exposed to direct solar radiation or when working in the sun during hot weather. This is due to that their bodies absorb a great deal of solar radiation because of their dark skin and sparse coat or hair, in addition to that they possess a less efficient evaporative cooling system due to their rather poor sweating ability as compared to cattle. Buffalo's skin has one sixth of the density of sweat glands that cattle skin has, so buffaloes dissipate heat poorly by sweating.

Management of Buffaloes

Buffaloes are managed under very different conditions over the world. These depend both on the geographic situations and for what purpose the buffaloes are used. Any system from multi-purpose animals kept in backyards to high yielding milk producers in advanced farms exists. Ninety-nine percent of the milk producing buffaloes are owned by small to medium land holding farmers and are merely a source of a small extra income. One or two buffaloes are a fairly common herd size managed under backyard system. They are fed crop resides like wheat straw, paddy straw, stovers, etc. This diet is sometimes supplemented with grazing and/or fodder by the cut-and-carry system and concentrates. In Punjab, the highest milk consumption in India is noted, approximately 800 grams per capita and day. Here, a well developed milk collection system is in operation through dairy cooperatives. The farms of this area are somewhat bigger than the national average, 10 to 15 buffaloes are common in a family farm. Although cows (crossbreds) are more common, almost all farmers keep a few buffaloes for house hold needs. Mixed farms are also common, where milk from both buffaloes and cattle are produced and sold commercially. At these bigger farms, all green fodder is produced at the farm and the buffaloes are fed a fairly balanced ration of green feed, concentrate and straw. Calves are neglected and often only used to secure milk let down. Female calves get more attention than male calves. The calf mortality is quite high in India. Milking is predominantly done twice a day by hand. Very few buffalo farmers have so far commenced machine milking, although the number is increasing. Larger farms, such as university and governmental farms contribute little to the total milk production of India. However, the yield per buffalo is higher than in other production systems. Some reasons for this are the better care and feed as well as a larger capital to invest in high yielding buffaloes by the university and government farms.

Temperament wise buffaloes are considered to be docile and friendly animals. Aggressive behaviour seldom occurs. For example, a new group of buffaloes is allowed to enter another groups' wallow or to drink from their water hole. Buffaloes are a bit dull and slow learners. It takes them a long time to get used to new situations and routines. However, they are easily stressed and may become very nervous if confronted with new situations. There are very strong bonds between the mother and her calf. If the calf is a male, the bond holds for approximately 2 to 3 years, thereafter the calf is driven from the clan. In the case of a female calf, the bond exists for life.

Traditionally, Buffalo have been fed poor quality forage as they have the ability to convert this to meat or milk. This fact should not be overlooked when formulating a diet. By increasing the quality of a ration will not necessarily be beneficial. As a rough guide a buffalo will consume 2.5% of its body weight as a daily dry matter intake. Feeding behaviour of buffaloes is restricted to grazers and only browses when feed is utterly scarce. Normally, buffaloes graze during the day. In case of extremely high ambient temperatures, grazing takes place in the morning and sometimes during night time. Buffaloes graze more and better than cattle. Thereby they consume more feed and nutrients per kg of body weight than cattle do. Newborn calves suckle their mothers within two hours of birth. Normal suckling frequency is approximately 6 to 8 times per day. The calves start to nibble grass at

3 to 4 weeks of age. When the calves have reached two months of age, forage starts to become more important than previously and soon most of the nutrient intake comes from forage rather than milk. Natural weaning of calves is usually within a year or before its mother's next parturition. The calf walks with its mother as soon as it can stand. This behaviour is different from cattle where the cow leaves its calf to go grazing.

The adult animals are extremely disease resistant. Calves especially rarely suffer from pneumonia or non-nutritional scours. The incidence of Lameness and Clinical Mastitis in adults is very less as compared to the cows. On the other hand, Vaginal and uterine prolapse, are the only major health problem effecting buffaloes. The reasons for this are unknown. Another complication arises is uterine torsion which most often occurs due to wallowing behaviour of adult females during their late pregnancy. Longevity is an important feature of Buffalo. Even in large herds (300+) it is not uncommon to find animals well over 20 years old still being milked next to freshly calved heifers.

Dehorning is not recommended as the horns provide a mechanism for body heat loss and a major role in breed identification.

Housing, provided it is well constructed, can be loose yard or cubicles.

Although buffaloes can adapt to harsh environments and live on low quality forage, its reproductive efficiency is often compromised by such conditions. Poor nutrition usually related to seasonal fluctuations in availability and quality of feed, delays puberty, age at calving and increases the duration of post partum anoestrous and long calving interval. In males, this results in loss of libido. Heat detection in Buffalo can be very difficult. For that reason artificial insemination is less successful in villages. However, the technique is similar to cattle. The success rates with artificial insemination are quite variable and vary according to season, in winter the rates are high as compared to summers in general. In field conditions, livestock keepers prefer natural service to artificial insemination.

Wallowing Behaviour

It is characteristic of Indian buffaloes. Wallowing has two purposes; the most obvious is that of cooling, the other is protection from insects. Wallowing during daytime is done during the hottest hours. Wallowing during night time is instead a way for the animal to protect itself from insects. The buffalo has few sweat glands and a dark skin which makes it difficult for them to maintain their body temperature in hot and humid environments. Wallowing is a very important way for the buffaloes to maintain their body temperature. When buffaloes enter the water, they defecate and/or urinate. This seems to be a way to mark their wallow. Wallowing behaviour is a learnt behaviour. While Buffalo love to wallow in water it is not necessary to provide them with a facility to do so. An animal that has not wallowed from birth with other animals will not do so on its own. If no water or mud hole is available, the buffaloes behave more like cattle. They will seek shade and graze more during the cool hours and less during the hot hours of the day.

Performance of Buffaloes

River buffalo are primarily a triple purpose animal producing milk, meat and draught power, whilst the swamp buffalo are kept in the main for meat and draught. The minimum levels of production and reproduction traits, which determine the productivity of the animal and shall make the system sustainable under optimal agro-climatic conditions (refer table 5 of Chapter 1). Milk yield from an individual influences the profitability and economic returns most. Some of the other important factors, which influence sustainability and economic returns, are age at calving, days open, days dry, lactation length and calving interval etc. System of management also significantly causes varied productivity in animals besides environmental influences. The expected genetic changes per unit of time for most of the economic traits is so meager that productivity and sustainability of a species generally depends upon genetic potential of the animal, environmental conditions and managemental variations. For a production system to be sustainable, buffalo rearing has to be operated as commercial dairy and must emphasize on more milk of quality composition.

The following are the salient points about the performance potential of river buffalo.

Milk

River buffalo holds the greatest promise and potential for production. The production of buffalo milk in the Asian-Pacific region exceeds 45 million tonnes annually of which over 30 million tonnes are produced in India alone. With selective breeding, improved management and the establishment of more dairy herds, milk yields are increasing worldwide. The individual 3,000 litre-per-lactation female, considered a record 30 years ago, is now common. There are many which yield 4,000 litres in a lactation of 300 days. Some have even attained a record of 5,000 litres. The potential for increased milk production, therefore, exists. The high-yielding Murrah is the Holstein-Friesian of the buffalo world. There is no reason why the Murrah and other riverine breeds should not exceed the volumetric yield of the best milk cows. Buffalo milk contains about twice as much butterfat as cow milk. There is no physiological need for concentrate feed to maintain this level. When concentrates are fed, the fat content rises. When fed adlib, it can exceed 15 percent. Fed to excess, this figure can be much higher. The buffalo, in fact, releases unwanted fat in the milk and stores only a minimum in the body tissues.

Average 305 days lactation

Milk production in some of the well-defined breeds of buffaloes in India varies from 1384 kg to 2033 kg . However some of the high producing buffaloes produce as high as 4000 kg and more in 305 days lactation yield, but their number is limited. In India, about 75 - 80 % of the non descript buffaloes do not come under any well defined breed and their average productivity is much lower than the buffaloes of milch breeds available in the country (2.61 kg vs 5.6 kg respectively). Breeding policy for their improvement in milk production have been formulated in various states but there is large gap in their implementation and animals are left to sporadic use of non selected and poor quality bulls for breeding. Our immediate target should be to double the productivity of these animals through systematic breeding by the use of superior germplasm.

Age at first calving

Fast growing heifers reach their maturity earlier than slow growing animals. This may be due to either heritable factor of fast growth rate or due to intensive feeding during the growth period. There is positive correlation between age at first calving and the length of their 3 calving intervals, thereby reducing the total productive period. Early calvers, yield less during first lactation since udder development takes place during pregnancy and at least 2 to 3 pregnancies are needed to attain full udder capacity. The milk yield during second lactation increases, with increasing age at which animal calved even if calving interval is kept constant. From third lactation onward milk yield is not influenced by age but from genetic potential and management.

Service period

Service period and dry period are interdependent factors and determine the length of productive days. The higher service period tend to reduce productive days and make the system less sustainable. Pregnancy depress milk yield, especially after 150 days of gestation. Service period is an important factor for sustainable production since longer service period is likely to leave the animal unproductive for longer duration and result into low profitability.

Dry period

Dry period is highly correlated with the service period and calving interval, a short calving interval is bound to have short dry period. Whereas longer dry period is likely to have longer service period and longer calving interval. Days dry prior to calving does not greatly influence lactation yield, unless the dry period is shorter than 30 days or so. Due to the intra animal correlation between dry periods over various lactations, it is difficult to estimate the effect of dry period on milk yield. However, shorter dry period gives more productivity as compared to longer dry period. The shortest dry period has been reported from buffaloes maintained at NDRI, Karnal (104 days) which are at reasonably acceptable level. At all other organised herds the dry period reported is longer.

Buffalo vs. Cow Milk

Although, both cow and buffalo milk are different with regard to their percent compositions but there is practically no difference in the nutritive value and digestibility of milk and milk products obtained from cow and buffalo milks.

Comparing the milk composition percent of buffalo milk with cow milk we notice that buffalo milk seems to be superior and better suited than cow milk.

1. It will be seen that buffalo milk has about 11.42 percent higher protein than cow milk. Protein Efficiency Ratio (PER) value of buffalo milk proteins to be 2.74 and that of cow milk as 2.49. Significant lower level of cholesterol content in buffalo milk is preferred as compared to the cow milk (Table 6).
2. Buffalo milk is also superior to cow milk in terms of important minerals, namely calcium, iron and phosphorus which are higher

by 92 percent, 37.7 percent and 118 percent respectively, than those present in cow milk.

3. Buffalo metabolizes all the carotene into vitamin-A, which is passed on to milk as such leading to more of vitamin-A as compared to cattle's.
4. Proteins of buffalo milk, particularly the whey proteins, are more resistant to heat denaturation as compared to the cow milk proteins. Dried milk products prepared from buffalo milk exhibit higher levels of un-denatured proteins when processed under similar conditions. In general, the reconstitution behaviour of dried milk products made from buffalo milk is indistinguishable from those made from cow milk. However, dried buffalo milk may be preferred over dried cow milk for those technological applications where higher levels of un-denatured whey proteins would be more desirable.
5. Cheese made from buffalo milk displays typical body and textural characteristics. More specifically, where chewing and stringing properties are specially desired as in the case of Mozzarella cheese, buffalo milk is technologically preferable over cow milk. In Italy, recently legislation has been introduced to restrict use of term "Mozzarella" only to those products exclusively made from buffalo milk (without admixture with cow milk). Certain traditional cheese varieties, such as paneer in India or pickled cheeses from the Middle-East countries, are best made from buffalo milk. Please note that cheese is a fermented product while paneer is a coagulated product from milk.
6. The presence of higher levels of various bio-protective factors, such as immunoglobulins, lactoferrin, lysozyme, lactoperoxidase as well as bifidogenic factors, render buffalo milk more suitable than cow milk for the preparation of a wide range of special dietary and health foods.
7. Buffalo milk is a rich source of Riboflavin and Vitamin-B_{12}. Vitamin-C and Thiamine are also found in good amounts. Small amounts of Folate, Pantothenic Acid, Vitamin-B_6 and Niacin are also found in Buffalo milk.
8. Calorie Content of Buffalo's Milk: 100g of Buffalo's Milk has 97 calories. Calories from fat are 61.
9. Health Benefits of Buffalo's Milk: Buffalo's milk is good for healthy bones, dental health, cardiovascular health, weight gain, etc.

The comparison between buffalo milk and cow milk is summarized in Table 6.

Table 6: Comparison of Buffalo Vs. Cow Milk

	Buffalo	Cows
Fat %	8.0%	3.9%
Protein %	4.5%	3.3%
α s1 Casein	1.44 -1.80 g/100ml	1.0 g/100ml
α s2 Casein	0.22 -0.28 g/100ml	0.26 g/100ml
β Casein	1.26 -1.58 g/100ml	0.93 g/100ml
κ Casein	0.43 -0.54 g/100ml	0.33 g/100ml
β lactoglobulin	0.39 g/100ml	0.32 g/100ml
α lactalbumin	0.14 g/100ml	0.12 g/100ml
Lactoferrin	5-32 g/100ml	5-28 g/100ml
Immunoglobulins	9.80 g/100ml	38 g/100ml
Cholesterol	8mg	14mg
Colour	Pure white	Creamy
Texture	Smooth	Less smooth
Taste	Sweet	Salty
Cell Counts	Very low	Higher
Average Yield/Lactation	1850 Kg	5500 Kg

Buffalo milk is commercially more viable than cow milk for the manufacture of fat-based and SNF-based milk products, such as butter, ghee and milk powders because of its lower water content and higher fat content. Most significantly, the lower cholesterol value should make it more popular in the health conscious market. By the virtue of greater opacity of casein micelles, coupled with higher levels of colloidal proteins, calcium and phosphorus, buffalo milk is more densely white and has superior whitening properties as compared to cow milk. Therefore, unlike the cow milk (which is pale-cream yellow in colour) and cow milk fat (which is golden yellow in colour), buffalo milk is distinctively whiter. UHT-processed buffalo milk and cream are intrinsically whiter and more viscous than their cow milk counterparts, because of conversion of greater levels of calcium and phosphorus into the colloidal form. Buffalo milk is, therefore, more aptly suitable for the production of tea and coffee whiteners than cow milk. Higher innate levels of proteins and fat render buffalo milk a more economical alternative to cow milk for the production of casein, whey protein concentrates and a wide range of the fat-rich dairy products.

Products from Buffalo Milk

Depending on its inherent qualitative and quantitative characteristics, buffalo milk is eminently suitable for certain types of both western products and region specific indigenous milk products in India. The innate qualities of buffalo milk such as high total solids content, higher fat percent, superior whiteness and viscosity, give buffalo milk preference over cow milk for the preparation of several Indian dairy products as well as specialised cheeses. From a technological point of view, buffalo milk can provide a wide variety of products: butter, butter oil, soft and hard cheeses, condensed or evaporated milks, ice cream, yoghurt and butter milk. The high fat content of buffalo milk makes it highly suitable for processing. For example, the production of one kg of butter requires 14 kg of cow milk against 10 kg of buffalo milk. Higher fat, total solid contents combined with the higher melting point of buffalo milk fat offers distinct advantages, which is well suited for summer climates. In many countries, buffalo milk is used to make traditional cheeses, for example, Mozzarella and Ricotta in Italy, Gemir in Iraq, Paneer in India, Domiati in Egypt, Pecorino in Bulgaria and Pickled cheeses from the Middle Eastern countries. There are also health issues if cow milk is used for milk products, because many people are allergic to cow milk and therefore, buffalo or goat milk is their choice for products, for milk, cheese and yoghurt etc. All the products that are made from cow milk can be made from buffalo milk with better keeping quality and high shelf life. The dairy processing units whether in private or organized sector both pay a premium prices for buffalo milk. The technologies have also been developed to offer solutions to the problems encountered in making products from buffalo milk.

Meat Potential of Buffaloes

Recent investigations and trials have convincingly established the fact that when a buffalo is fed well and managed for early slaughter (at a live weight of 350 to 450 kg), a yield of palatable, high-grade meat can be obtained at a competitive cost. A buffalo carcass has less fat, less bone and a higher proportion of muscle than a cattle carcass. Furthermore, a buffalo carcass does not need the customary and costly chill hanging for a specific period to ensure the "eatability" of the product.

It is well known that the buffalo is remarkable for its feed conversion ability; the young buffalo calf achieves a daily weight gain of 800 grams without any supplement feed. Similarly, the power of the full-grown work-buffalo does not come from high level nutrients. Such abilities of these magnificent animals should be utilised to the maximum especially in the country like India. They are well suitable and sustainable in India where animals are reared on low input system.

Buffalo Meat vs. Beef

Buffalo meat, like the milk is lower in cholesterol and higher in mineral content than that from cows. Lean buffalo meat has less than half (44%) the total fat content of lean beef and has proportionately less saturated fat. In addition to these obvious health benefits there is the added bonus for those who are concerned about Bovine Spongiform Encephalitis (BSE) there has been no incidence of BSE in any buffalo anywhere in the world. When cooked there is little noticeable difference in the two meats, either visually or in taste or texture. The optimum weights for slaughtering young bulls appear to range between 400 and 600 kg which are attained at 25-30 months of age. Care should however be taken to ensure that finished animals are not over fat, so there is no marbling as it is unwanted and time consuming to remove from the carcass.

Draft Power

Buffaloes are laboured intensive and cost-effective. They are the most versatile of all work animals in the variety of tasks which they can be taught to undertake. All buffalo breeds have a strong milk/ meat entity. However, presently the output of thousands of buffaloes is in the form of work energy rather than the direct provision of food as milk or meat. In economic terms, buffaloes are productive and efficient, especially in those agricultural countries where there is ample manpower and little motivation for mechanization. In India, the mechanization in agriculture has lead to decrease in the use of animal power in the last couple of decades leading to slight decrease in the population of buffaloes as farmers are more interested in rearing the she buffaloes for milk. Buffaloes in most part of the country are primarily used for milk, spent buffaloes and young and emaciated males are sent to slaughter houses for meat production while adult males are generally used for work. Buffalo males of Murrah breed have efficient work capacity especially for load pulling and ploughing in rice cultivation. Though, buffalo bulls are slower in movement than cattle but they can pull heavier load and cover about 3.2 Km per hour as compared to 4.8 to 6.4 Km per hour by draft bullocks. Though large buffalo bulls are able to pull heavy load but are not maneuverable like small and compact animals and hence move slowly.

Animal Dung

Animals dung is also one of the major contributions towards economic contribution of livestock and account for about 10% of the value of output of livestock sector. The dung is used as organic manure and fuel as source of energy (bio gas). It estimated that annual compost production from three to four animals is 2230 kg and when passed through biogas plant, the quantity of Farm Yard Manure (FYM) produced is 2676 kg. FYM is a good source of Organic farming (Vermicompost Development). This definitely adds to the economic value of dairying if dung is used properly.

Summary

The river buffalo has 50 chromosomes and the swamp type has 48, the amount of genetic material is the same in both these types of buffaloes. Swamp buffaloes have a very low milk yield (approximately 400 -500 kg / lactation) and are not used as milk producers. Buffaloes are distributed from India through Pakistan into the Middle East and in Bulgaria, Hungary, Turkey, Italy and Egypt. Buffaloes are perfectly suited to the hot and humid climate with muddy terrain of India. Buffalo's skin has one sixth of the density of sweat glands that cattle skin has, so buffaloes dissipate heat poorly by sweating. The animal has a huge potential in dairy industry and is already a leading producer of milk in India. The buffalo milk is more suitable for milk products preparation due to its high butter fat percentage and high percent of solid not fats. The taste of buffalo milk is considered superior by Indians.

Chapter 3

Sustainable Breeding Policies for Dairy Cattle

Sustainable livestock production aims at maintaining economical, high yielding animals per unit of time and inputs on long-term basis with least deterioration in natural resources; which contribute life support system for agriculture and animal production and with least contribution to environmental degradation. For sustaining the large and burgeoning needs of fast growing human population, we have to work out strategies that are sustainable for a long time to come. The intensive use of input-based high yielding technologies in animal husbandry is required. Therefore, there is need to follow a multidimensional approach to enhance the productivity of indigenous cattle population by enhancing their genetic potential. Overall development in livestock sector comprises selecting the animal for overall adaptive value, production of superior males and dissemination of their semen in vast population, improving conception rate from AI under field condition and developing need based and economically viable technologies for small and marginal farmers.

Breeding Policy for Cattle

The major factors that determine the success or failure of breeding programme are relevance of the breeding programme for genetic improvement in the prevailing scenario, objectives/goals of the programme and breeding strategies. The relevance of the breeding programme is of paramount importance. Bringing about genetic improvement beyond the nutritional or disease-limited environment in which the population lives can be of disadvantage and often counter productive. It has been reported in literature that it makes little sense to initiate genetic improvement programmes in livestock populations where annual feed intake is less than 1.5 times the animals maintenance feed requirements. Rather, it was recommended that feed intake levels of the population should be twice the maintenance levels to launch any genetic improvement programme. Another important factor is the availability of sufficient infrastructure to support the programme like AI and veterinary services, procurement of milk and supply of other inputs at the doorsteps of the farmers. The breeding objectives must be relevant to the prevailing production system of an area. In tropical countries like India, the selection goal must identify animals that perform well under heat stress and can cope up with the seasonally available fodders/crop residues and have the resistance for diseases prevalent in tropical environment. It was suggested that for best efficiency the genotype must have an estimated producing ability about 30% above that possible with the projected feeding.

The goal of any programme should be to increase the genotypes of that animal population to best fit the total environment of its production system including climate, management, social and economic factors. It has been suggested that single purpose breeding goal should not be practiced, while multi-purpose breeding objectives should be considered for the overall 'adaptive value' of an individual in the population. The breeding strategy for the improvement of indigenous cattle (20%) had been selective breeding in well defined breeds and grading up of non-descript zebu cattle (80%) with the breeds available in the area or adjoining area. On the other hand, if the feed and fodder along with good infrastructure are available in plenty the cross breeding is judicious use of semen from exotic bulls on nondescript cattle is recommended. Past experiences have shown that the wrong implementation of crossbreeding in our native breeds at organized farms and under filed conditions has depleted the number of animals of these well-defined indigenous cattle breeds.

There is an urgent need to monitor strictly the on-going breed improvement programmes of various indigenous breeds at organized farms and to widen these programmes by net-working various organized herds and bringing farmers herds under the progeny testing programmes to increase the number of animals of well defined breeds. Further, our breeding policy intervention for sustained milk production should be situation specific and should not be generalized. The determinants of recommending the breeding policy are description of production system, definition of breeding objectives (trait or combination of traits) and knowledge of breed resources (genetic variation in production and adaptive traits). Unfortunately, the production systems vary considerably from area to area and even from farmer to farmer. Data on performance traits, adaptive traits and unique traits like resistance to a particular disease/parasite and tolerance to environmental stresses are absent or inadequate for most of the breeds.

Under the prevailing circumstances, a short term interim breeding policy is proposed:

- Conservation and improvement of well defined indigenous dairy and dual purpose breeds of cattle in their respective breeding tracts.
- Pure breeding in draft breeds with emphasis on both draft power and milk production.
- Grading up of non-descript animals with locally adapted dairy/ dual purpose breeds, if the farmers opt for.
- Crossbreeding of non-descript cattle in urban and semi-urban areas, and as per farmers choice and resource availability in other areas; the selection of exotic breed (HF or Jersey) depending upon the availability of resources and market demand for milk.
- Inter-se mating amongst crossbreds to contain exotic blood between 50-75%.
- It is advocated that serious efforts should be made to identify breeding males of high genetic merit and their effective use under field AI programme. The choice of the farmer for selection of quality semen for his animals should be given priority.

Grading Up of Local Nondescript Cattle

Large number of low producing nondescript cows (80% of cattle population) is one of the major constraints to launch any effective breed improvement programme. These nondescript animals have poor breeding efficiency leading to longer inter calving periods coupled with shorter lactation length. As there is a declining trend in the number of indigenous cows over the years, it is desired to curtail the number of unproductive / low productive nondescript cattle.

The nondescript cattle can be genetically improved by grading up using either elite pedigreed or progeny tested bulls of well known indigenous breeds like Sahiwal, Tharparkar, Red Sindhi, Gir, Deoni, Hariana, Ongole, Kankrej etc. available in their breeding tract. These breeds are maintained on several organized government and non-government farms. The bulls to be used for this purpose should be produced from superior dams which have more than 2000 kg as lactation yield for milch breed of Sahiwal, Tharparkar and Gir cattle and more than 1500 kg for dual type cattle breeds of Hariana, Kankrej and Ongole etc. This will improve the milk yield by 500 to 800 kg in the first generation. By use of zebu bulls of high transmitting ability for grading up subsequently in initial generations, milk yield can be improved to the extent of 5 to 10 percent per annum. In a period of 5-6 generations of continuous grading up, the nondescript stock will be transformed into well-defined purebreds.

Improvement of Indigenous Cattle by Selective Breeding

Superior bulls of Sahiwal and Red Sindhi are required for multiplication of their high quality germplasm for enhancing the productivity of vast indigenous cattle's. It is necessary to take large scale genetic improvement programme for indigenous breeds in their breeding tract through selective breeding. The existing herd of milk and dual purpose breeds like Sahiwal, Gir, Red Sindhi, Tharparkar, Deoni, Ongole and Haryana needs to be strengthening further. By selective breeding, it is expected the genetic improvement will be achieved ranging from 1-1.5% per annum in herds at organized farms and 8-10% per annum in farmer herds (Stonker, 1953). Such breed improvement programme must be evaluated from coordinating bodies for selection of young bulls and testing their genetic merits at organized herds and as well as under field conditions. The progeny testing is the best basis of selection for bulls to be used for selective breeding.

Evaluation of Indigenous Breeding Bulls of Cattle

Traditionally, pedigree selection and progeny testing are practiced for identification and evaluation of breeding bulls, newer techniques like Harvey's least square, Regressed least squares, Best linear unbiased prediction (BLUP) and REML techniques are now being used for estimating breeding values of sire. A project on associate herd progeny testing of Sahiwal bull was initiated in 1978 at NDRI karnal, with cooperating farms which are government livestock farms NABHA (Punjab), Durg (Chattisgarh), Lucknow (UP), Hisar (Haryana), CSAUAT (Kanpur, UP), GBPUAT, Pantnagar (Uttaranchal), Military Dairy Farm, Meerut (UP) participating voluntary for testing the Sahiwal bull the breedable females of their herd. However majority of small herds belonging to different organization discontinued in between due to physical and local administrative constraints. Test bulls are put to test mating under a set and till now various sets of Sahiwal bulls have been put to progeny testing and bulls have been evaluated (for detailed progeny testing programme layout refer Chapter 5). Top ranking Sahiwal bulls out of whole set of test bulls are then used in the breed improvement programme and being used for Elite mating for production of young bulls for future breeding.

This is one example similarly other cattle breeds should also be included and a network project on cattle is needed collaborating the various associated and organised herds.

Crossbreeding of Nondescript Cattle

Cross breeding of nondescript cattle as per farmer choice by using exotic breeds like Holstein Friesian in plain areas of the country or with Jersey breed which is better suited to hilly areas of the country (due to its smaller body size of Jersey cattle and less body weight as compared to the Holstein) is recommended. Crossbreeding can produce more milk but they need more and better feed and resources to exploit the genetic potential of exotic breeds. Crossbreeding of nondescript cattle in urban and semi urban areas is advocated to ensure adequate milk supply where there are facilities for rearing and maintenance of high yielding cattle are available. A number of studies have been carried out to analyze the performance of cross breeds.

Studies have shown that age at first calving in crossbred (27-30 months) is much lower than the indigenous breeds. Average lactation yield in crossbred varied from 1100 to 4000 kgs. The superior performances of crossbred over indigenous breeds have been generally explained due to non additive gene effects (Hetrosis). The result of various cross breeding studies shows that the optimum level of inheritance was between 1/2 and 5/8. Beyond this level, the overall adaptability of cross breed gradually decline, resulting in prolonged first calving and also decline in lactation yield. Only the nondescript low producing animal should be breed through cross breeding.

Future Breeding Strategies for Cattle

Keeping in view the present scenario of livestock production prevailing in our country, it is suggested that

1. Selective breeding in well defined milch and dual purpose breeds has to be followed.
2. Well defined indigenous breeds should not be subjected to crossbreeding
3. There is an urgent need to conduct breed-wise census for systematic planning of any breed improvement programme. Breed Associations and Breed Societies may be formed in the breeding tract of various indigenous breeds.
4. The surplus draft animals can be graded-up with dual purpose breeds available in that area or adjoining area.
5. Improve conception rate from AI under field conditions although National Policy of Cattle and Buffalo Breeding (NPCBB) which started from year 2000 for a period of ten years is aiming at door step delivery of Artificial Insemination (AI) to the farmers and to cover all the breedable cattle by AI by 2010. NPCBB is now extended upto 2015.
6. Strengthening various sire evaluation programmes at organized herds and under field conditions using farmers herd. Implement aggressive extension strategies for awareness regarding importance, utility of progeny testing programmes. Surplus semen doses from ongoing sire evaluation programmes to be given to developmental agencies for dissemination under field conditions. Information on available germplasm (semen) with various agencies should be published in the form of sire directory.
7. Bull mother farms for production of superior bulls / semen or MOET nucleus herds of elite cows of different breeds should be established in their respective breeding tracts to produce more males of higher genetic merit.
8. Emphasis should be given to selection the animals for overall adaptive value instead of single trait selection on milk yield.

9. The different SAU's and Regional Stations should demarcate different areas and the research should be done to solve the problems faced by the farmers. Further, a good deal of interaction and cooperation should be there among research and extension workers of SAU's and State Animal Husbandry Departments.
10. The nondescript and scrub bulls should be castrated. Clear cut culling policy for surplus unproductive animals to down size their number systematically over the years needs to be framed.
11. The developmental technologies should address the specific needs of the farmers. These technologies should be easy to use, feasible, economically viable, self-sustainable and eco-friendly under farmers conditions.
12. More emphasis should be given to quality feed production, multiplication, processing and marketing facilities for the areas under genetic / breed improvement programmes.
13. Different methods of providing optimal environment to animal especially during summer months in crossbred cows. Like mist cooling should be evolved to relieve heat stress, (for details refer Chapter 15).
14. Strongly educate farming community about the importance of draft animal power and campaign to save them.
15. Long term planning has to be initiated to change traditional animal production system gradually to commercial one in the areas having requisite resources to support the dairy enterprise.

□□□

Chapter 4

Sustainable Breeding Policies for Dairy Buffaloes

Today, India has achieved the distinction of being the top milk producing country in the world. The trend of milk production in our country indicates that of the total milk production in the country about 53.64% is produced by buffaloes. About 95 percent of the total milk handled by various dairy plants in the country comes from buffaloes.

Buffalo Breeding Policy

The current breeding policy recommended by the National Commission on Agriculture and adopted by the Central and State Governments was based on the important considerations which were similar to those of the scientific panel on Animal Husbandry. The major features of present buffalo breeding policy in organized herds are as follows:

A) Improving Buffaloes by Selective Breeding

For effective selective breeding, the existing organized farms of Murrah, Surti, Mehsana, Nili Ravi, Nagpuri, Bhadawari and Jaffarabadi should be strengthened for production of breeding bulls.

In certain pockets of states like Gujrat, Rajasthan and Karnataka, "Surti" is recommended to be breed of choice; Nili Ravi and Murrah are the breed of choice in the Punjab, Haryana, Western U.P and other parts of the country where sufficient feed and fodder resources and good market for milk are available. In Assam, "Swamp buffalo" is also recommended for breed improvement programme. Majority of superior animals of well defined buffalo breeds are maintained at various organised farms and there is an urgent need to conserve and improve the production of superior germplasm and its dissemination. A lot of early work had been carried out by different workers for genetic improvement of buffaloes. However, there is a paucity of information on the systematic evaluation of the breeding worth of buffalo bulls, especially, for efficient milk production. A systematic effort on progeny testing programme of buffaloes has been started by the Indian Council of Agricultural Research (ICAR) through all India Coordinated Research Project on buffaloes since 1970-71. The thrust area was to evaluate the sires with a view to produce proven buffalo bulls for enhancing milk production. The efforts made through this venture have been able to standardize the testing methodology and germplasm evaluation.

The infrastructures have been created which are capable to generate germplasm in the form of bulls and frozen semen at some of the testing centres. During the 8th Five Year Plan, efforts have been made by coordinating and collaborating with various Institutions engaged in buffalo research and organised buffalo farms to meet the objectives of producing large number of genetically superior buffalo breeding bulls on the basis of their pedigree and progeny performance under the Network Programme of Buffaloes. Use of semen of different bulls of a set in multiple herds enhances the accuracy of selection of bulls under progeny testing programme by overcoming the constraint of smaller herd size and producing more number of daughters per bull. At present there are seven institutional herds (CIRB, PAU, NDRI, CCSHAU, IVRI, CCBF and FAIZABAD) involved under the Network Programme of Buffaloes in Murrah breed and one institution each involved in Network Programme of Nili Ravi (CIRB, Nabha), Jaffarabadi (GAU, Junagarh), Surti (MPUAT, Vallabhnagar), Badhawari (IGFRI, Jhansi), Pandharpuri (MPKV, Kolapur), Godavari (ANGRAU, Venkataramangudem) and Swamp breed (AAU, Khanapara).

Strategy for Selective Breeding

Identification of proven sires

To start a progeny testing programme there is a need to have the proven bulls or their semen for the nominated mating. The progeny tested bulls / semen can be procured by identifying the sires at other institutional herds which are the part of the associated herd progeny testing programme.

Selection of elite buffaloes

Elite buffaloes are to be selected based on either of the following criteria and nominated matings are to be done with the proven bulls available in the herd.

Best lactation yield	=	2500 Kg
First lactation yield	=	2300 Kg
Breeding Value	=	1950 Kg

Selection of test bulls

All males born due to nominated mating in multiple herds are to be reserved for the subsequent selection for test mating based on the following criteria:

Dams best lactation 305 days yield	=	2500 Kg
Breeding Value	=	1950 Kg
Semen Freezability		
Breed Characteristics		
Son of test/ proven sires		

Test bulls should be healthy and free from zoonotic diseases

Freezing and storage of quality semen of test bulls

About 6000 quality semen doses from each test bull are to be frozen for test mating and store till the evaluation of bulls being completed in multiple herds. For conducting the field progeny testing programme, the effort has also been made for improving Murrah breed and for that it has been estimated that about 3000 - 4000 additional doses from each test bull are to be frozen for test mating under field condition.

Test mating programme

Breedable buffaloes in the multiple herds as well as in the field are to be identified as test herds and shall be inseminated with the frozen semen of atleast 15 test bulls over a test cycle of 18 months period (for detailed progeny testing programme layout refer Chapter 5).

Recording of information and data

The information on growth, reproduction, production performance; age group wise mortality of buffaloes; bull wise semen production and utilization and daughters born shall be maintained in the herd.

Buffalo Bull Evaluation

Sires can be evaluated on the basis of their daughters 305 day milk yield in multiple herds using the method as under:

$$\text{Sire Index} = \mu + \frac{2n}{n+12}(D - CD)$$

Where,

μ = Overall average of daughters 305 days first lactation milk yield.

n = No. of daughters / sire.

D = Daughters average 305 days first lactation milk yield and

CD = Contemporary daughters (over all the herds) average 305 days first lactation milk yield.

All the sires will be ranked on the basis of above sire index of which 20-30% top ranking bulls shall be selected for future use as proven bulls for production of superior germplasm. This can ensure sustained maintenance and production of improved germplasm on a large scale for use in buffalo improvement programme and for establishing linkages with other institutions in India.

B) Improvement of Buffaloes by Grading Up

The low producing buffaloes in areas where feed and fodder resources are not available in sufficient quantity, can be genetically improved by upgrading with superior established breed of buffaloes. The grading up of with superior germplasm from bull with high genetic potential would enhance milk yield in next generation. The programme may also help to increase about 2-3 times the expected milk yield within two generations in nondescript buffaloes which is about 60% of the whole buffalo population in the country.

Benefits of the Sustainable Buffalo Breeding Policies

1. The methodology developed for sire and cow evaluation under organized herds following sustainable breeding programme can be used by different R&D organizations for undertaking large scale progeny testing programme.
2. The surplus semen of superior and proven bulls can be given to organised herds or other R&D organisations for production of young superior males of higher genetic merit for production of future young males.
3. Genetic improvement of 1.0 to 1.5% in milk yield per annum can be obtained at organised farms and about 5.0 to 10.0% under farmer herds conditions.
4. Supply of best superior male germplasm in rural areas will enhance milk yield and help in the development of farmer's herds and establishing linkages with dairy development agencies and more institutions in India.
5. May help to generate more economic return by selling the best semen of bulls from organised herd and producing superior male and female germplasm in the rural households and thereby sustain the breeding programme.

Future Buffalo Breeding Policy

Over the years it has been observed that for continuing the buffalo breed improvement programme the present breeding policy where the prime objective is to improve the milk production, may be modified for which the following areas could be explored for effective and more sustaining the breeding programme

1. The milk processing traits may be incorporated for sustainable buffalo breeding programme keeping in view of more demand of buffalo milk by the dairy industry.
2. ONBS programme may be adopted with or without MOET for higher genetic improvement in buffalo.
3. The superior buffalo germplasm could be multiplied with the adoption of MOET in the breeding programme.
4. The methodology for evaluation of sires may be modified by giving importance to Test Day evaluation rather than 305 days milk yield of daughters.

5. Research areas are also to be strengthened and explored so that the results of other reproductive bio-techniques like IVF, polymorphism study with respect to various economic traits, sexing of embryos, transgenic approaches etc. could be incorporated in the breeding strategy for bringing about rapid genetic progress in dairy buffaloes in future.

□□□

Chapter 5

Progeny Testing Programme and Sire Evaluation

India has a very rich reservoir of animal genetic resources and possesses large number of diverse cattle and buffaloes breeds. Milk production cannot be improved by increasing the number of cattle; the other option is to increase the average productivity by improving the genetic merit of the dairy animals. Therefore, to improve the genetic potential for per animal productivity it is important to identify and select the superior animals with high genetic merit for milk production throughs various breeding programmes. Genetic improvement can be brought about by judicious selection of both sires and dams. Sire selection is more effective and progeny testing is an excellent basis for sire selection.

History

1. Approximately 2000 years ago : Roman Varro, first advocated a compendious idea of P.T.
2. In 18th century : Robert Backwell used the P.T. for assessing the worth of animals
3. In 1840s' : Danish farmers evaluated their sires through progeny testing.

P.T. in India

1. In 1964-65 : Buffalo improvement project was initiated at GLF, Hisar, (India) for selecting bull through P.T.
2. In 1970-71 : P.T. programme for buffaloes has been started by ICAR system through AICRP on buffalo
3. During VIII plan : Field P.T. for crossbred bulls in PDC, Meerut has been approved by the ICAR.
4. During IX plan : P.T. of Murrah buffaloes under field condition (around CIRB, PAU & NDRI) has been approved by the ICAR.

Reasons for Sire Selection

It is well established that the major part of the genetic improvement in monoparous animals like cattle and buffaloes is due to extensive use of males with high genetic potentials. The genetic potential of a male animal for sex limited traits like milk production could reliably by estimated only from its progeny performance and moreso when the progeny is reared under different managemental conditions (different herds).

Selection of sire rather than of the dam is very important. This is because of the followings reasons

1. A sire can leave over-whelmingly largenumber of progeny as compared to female. For example, whereas a she buffalo may produce ten to twelve progeny in her lifetime. A bull may beget about 50 progeny each year for about ten years through natural service and this number can be multiplied a thousand fold or more by using Artificial Insemination Technique or Embryo Transfer Technology.
2. The sire is "more than half of the herd", this because of approximately 61% genetic improvement (Balain, 1988) is acquired through the sire path.
3. Only few sires are needed for breeding a herd and hence intense selection can be made among males than females. This is because most of the female progeny born are selected to replace the old and unproductive cows & buffaloes only. As a result of higher intensity of selection in case of males, the selection differential is larger.

4. Sire selection is not only important within the herd or breed but is also equally important in grading up and crossbreeding programmes. The male progeny produced by better dams mated to best sire, are selected for breed improvement programmes.

Progeny Testing Programme

- Most important and best basis of selection
- Individuals are evaluated based on their progenies production performance
- Progenies give the best and most reliable information about the genetic merit of their parents
- P.T. Provides the true estimate of breeding value of an animal
- Progeny testing also gives more reliable or accurate estimate of breeding value with the increase in progeny size
- Progeny testing programme for buffaloes under field condition has been developed with the main aim that the bulls are to be evaluated based on their progenies performance raised in farmer's herds. Secondly, the germplasm of genetically superior proven bulls has to be conserved and multiplied for dissemination further to farmers and developmental agencies for improvement of buffaloes.
- Points to be considered in the Progeny Testing Programme
  - Number of bulls put under test
  - Number of daughters per bull tested
  - Proportion of population mated to test bulls
  - Proportion of daughters available from cows allotted per bull
  - Total population size

Prerequisites for progeny testing programme

- It is always better that more than required number of males is kept for progeny testing.
- The facilities should exist for frozen semen, and to record information on progeny. Moreover, a large breedable famale population is required.

Types of Progeny Testing Scheme

Single herd progeny testing scheme

Individuals are evaluated based on their progenies production performance raised under single organized herd level. In this method small progeny size (3 to 4 daughters per sire) are used for evaluation of sire's which leads to lower accuracy.

Associated herd progeny testing scheme

Most of the institutional herds, where progeny testing programme were undertaken had small herds of 150 to 300 animals. This is why small batches of 4 to 8 bulls were evaluated in each set and ranked on small progeny size from 3 to 4 daughters per bull. Consequently, associated progeny testing was initiated to increase the population size and to obtain additional number of daughters per bull for evaluation of sires. Also Progenies raised in different environmental condition leads to more accuracy in evaluation of sires, since genotype environment interaction is taken into account.

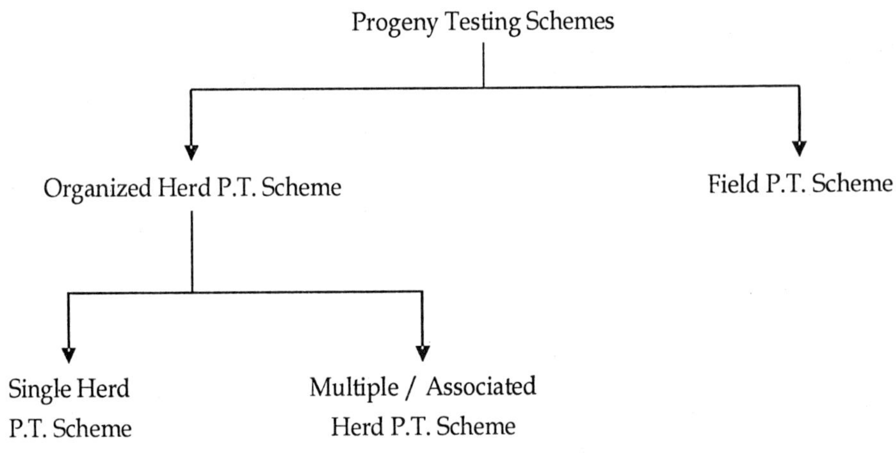

Methodology for Progeny Testing Programme

Identification of progeny tested proven bulls

The bulls which were earlier progeny tested and found proven at main centre or other institutional herds are also to be identified and initiative is to be taken to procure and store the quality semen doses of progeny tested proven bulls through an inter institution linkage for continuing the progeny testing programme in other herd or field.

Selection of elite animals for nominated mating

The criterion for selection of elite animals depends on superiority of the animals. The elite heifers would be selected initially and periodically in the operation area, could be multiplied, if possible, with the adoption of multiple ovulation embryo transfer technology (MOET) in the breeding programme. Elite heifers are to be selected based on either of the following criteria and nominated mating is to be done with the identified proven bulls.

- Breed characteristics of cows & buffaloes
- Heifers should be healthy, reproductively sound and free from zoonotic disease
- Dam's peak lactation yield and fat percentage, if available

Identification of test bulls

To begin a progeny testing programme, a minimum of ten test bulls are to be selected from the reserved males born due to nominated mating in institution herd or field based on all the following criterias

- Breed characteristics of bulls
- Bulls Expected Predicted Differences (EPD)
- Son of test or proven sires
- Age of the bulls do be put under P.T.
- Age at initiating training of the bulls
- Age at the first time of semen collection of the bulls
- Semen freezability traits
- Dam's age at first calving
- Dam's lactations 305 days milk yield
- Dam's average of lactations 305 days milk yield
- Dam's best lactation 305 days yield
- Dam's lactation wise calving interval
- Test bulls should be vaccinated against FMD, HS, BQ
- Test bulls should be healthy and free from TB, JD and other zoonotic diseases

Freezing and storage of quality semen from test bulls

For conducting progeny testing programme, the effort has to be made so that 3000-4000 doses of quality semen from each test bull are to be frozen for field condition in addition to doses of semen required for progeny testing programme (5000 doses) for main center or institutional herd.

Test mating programme

Breedable female animals in the herd or field are also be identified as test herds and shall be inseminated with the frozen semen of test bulls over a test cycle of 18 months period for buffaloes, 24 months for indigenous cattle and 30 months for crossbred cattle. For detailed information on test mating refer the progeny testing layout as explained in later part of this chapter.

Recording or generating and analysis of information / data

The information on growth, reproduction, production traits keeping in view of goal of the programme, other performance records, age group wise mortality of animals, bull wise semen production and utilization and daughters born shall be recorded, maintained and analyzed.

Performance Recording System

Animal performance recording is an essential prerequisite for deciding and planning breeding programmes for bringing desired genetic improvement. The herd performance recording system has immensely helped in improving the productivity of animals in dairy developed countries. Absence of systematic recording of performance of farmer herds is considered as major constraint for not augmenting the large scale dairy herd improvement and development programmes in the country. Increasing productivity through genetic improvement programmes requires identification of superior animals in the population to be the major contributors to the genetic material of future generation. Different performance recording systems being traditionally used at organized breeding herds could not simply be adopted on farmer's herds. So, several Research and Development Organizations initiated recording of the data on performance traits in farmers herd. NDDB started this system under Dairy Herd Improvement Programme (DIPA) in 1987 in Mehsana district. On the same line Kerala Livestock Development Board (KLDB), Central Herd Registration Scheme (CHRS) under Department of Animal Husbandry and Dairying, Ministry of Agriculture, Government of India and other institutions like BAIF, PAU and NDRI had also been established the performance recording systems with the objective, identification of superior germplasm through systemic milk recording and registration of the animals under field progeny testing programmes.

The milk recording will be done by providing two performance recording cards for each animal. One performance recording card for each animal should be given to trained milk recorder while the other performance recording card for the same animal should be given to each household. The recording of milk yield should be in equal interval may be based on daily / weekly / monthly basis and on the test day record should be made not only for milk yield but also for fat%, protein% and somatic cell score of each sample of milk.

Performance recording at organized herds

In general the performance data recorded on each animal individually identified along with pedigree information at organized herds are being utilized more judiciously in decision making for selection and mating of animals as well as in adoption of efficient management practices. However, these organized herds possess only a microscopic fraction of total cattle and buffalo population in the country and moreover being small sized herds, the overall genetic improvement at organized herds could not make significant mark on increasing the productivity of animals over the years. Efforts have been made to run breed improvement programmes especially in bull selection through progeny testing associating various organized herds for a specific breed group. On such line some dairy cattle improvement programme in the country such as on Sahiwal, Ongole, Hariana cattle and some Buffalo breeds have been initiated. Still such breed improvement programmes are on small scale and quite far from meeting the requirement of adequate number of superior bulls and their germplasm. For undertaking various cattle and buffalo herd improvement programmes including grading up a non-descript stock, improvement on well defined cattle and buffalo breeds and crossbred cattle population, huge number of quality breeding bulls are required. Therefore, in order to meet high requirement of bulls, there is need to enlarge the net working of associating organized herds and farmers herds along with strengthening AI facilities to cover more proportion of cattle and buffaloes for undertaking an accurate evaluation of large number of breeding bulls on the basis of performance of their progenies. The major constraint to carry out the selection of bulls under field condition is absence of identification of farmer's herds and performance recording system.

Different performance recording systems being traditionally used at organized breeding herds could not simply be adopted on farmer's herds. The field data recording methodology used in developed countries could also not be transferred as such, as it is impractical under our village condition. Since the dairy husbandry scenario in our country side representing the farmers with small sized herd of low producing non-descript cattle and buffalo following poor breeding, feeding and management practices and of inadequate marketing facilities for milk/milk products and animals, absence of farmers cooperative unions/societies, poor infrastructure of reads and communications, illiteracy amongst farmers etc have not made yet receptive environment to herd performance recording system under field conditions. Thus there is need to develop infrastructure facilities to under take performance recording under farmer's herd condition. For detailed field performance recording, refer Chapter 6.

Sire Evaluation

In the existing progeny testing programme, the Contemporary Comparison Method is the method of choice for genetic evaluation of test bulls for the main center or institutional herd.

Contemporary Comparison Method (Robertson and Rendel, 1950)

$$I = \mu + \frac{n}{(n+12)} \quad [D - CD]$$

Where,

I = Sire index

μ = Herd average

n = Number of daughters

12 = Constant based on sire error variance

D = Average of all daughters of a sire under test

CD = Average of contemporary daughters of the sire

For estimating the sire indices, the herd average along with first lactation 305 days milk record of daughters and their contemporaries are required for evaluation of test bulls more sustainably. However, non-availability of accurate 305 days first lactation milk record of daughters and contemporaries of a test bull in most of the time may make the sire evaluation programme non-sustainable under field condition. So, the Simple Daughter Average Method is the method of choice for genetic evaluation of test bulls for field conditions.

Simple Daughter Average Method (Edward, 1932)

$$I = D$$

Where,

D is the average of all daughters of a sire under test.

Today in most of the developed countries like USA, Canada, Netherlands etc. the use of test day milk yield model instead of conventional 305 day lactation milk yield model is of more interest for genetic evaluation by predicting daily / test day breeding values of animals with the help of polynomial functions using Random Regression Model where random regression coefficients are estimated using functions such as Wilmink's Function or Legendre Polynomial Function. The main advantages of using test day records under test day model for genetic evaluation of animals are, individual test date effects can be included in the model, number of records per animal and the interval between records can be accounted, avoid the necessity of extending short lactations thereby reducing the generation interval, increase the accuracy of adjustment for fixed effects leading to more accurate genetic evaluations and account for individual differences in the shape of lactation curves of animals which include the persistency of lactation.

All the test bulls will be ranked on the basis of average estimated total breeding values of their daughters of which 20-30% top ranking bulls shall be selected for future use as proven bulls for production and dissemination of superior germplasm to organized herds or other research and development organizations and in rural areas for the development of farmer's herds. Table (7) reviews the work done in India under the field progeny testing scheme.

Constraints of Progeny Testing Programme

1. Lack of financial resources and physical facilities do not allow production of large enough number of daughters from sufficient number of bulls simultaneously both in the farm and field conditions

Table 7: Research Work done in the field of Progeny Testing (India)

Types of P.T.	Location	Breed	References
Field	Anand	Surti Buffalo **22** (480)	Nagarcenkar & Rao (1983)
Field	Kerala	Crossbred **30** (1357)	Chacko *et al.* (1984)
Field	Maharastra	Jersey cross **32** (1026)	Chauhan *et al.* (1994)
Field	Maharastra	Jersey cross **32** (1026)	Godbole *et al.* (1994)
Field	Kerala	Crossbred **56** (2623)	Deb *et al.* (1998)
Field	HAU PAU BAIF	HF cross **30** (2700)	Kaushik & Garg (2000)
Field	Karnal	Crossbred	Gaur (2001)
Associated Herd	NDRI, Karnal GLF, Lucknow Vet. College, Durg	Sahiwal	NDRI, Annual Report (2004-05)
Associated Herd	CIRB NDRI PAU HAU	Murrah Buffalo	NDRI, Annual Report (2004-05)

Figures in bold and parentheses indicate number of bulls considered and number of total progenies, respectively.

2. All the present results of progeny testing suffer from the deficiency of adequate number of daughters per bull
3. Costly and requires large investment for quite long time
4. Illiteracy among the farmers
5. Compulsion of the farmers to sell their animals
6. Insufficient incentive to the farmers
7. Inaccurate information due to casual milk recorders
8. It increases the generation interval
9. Genetic gain per year is low
10. Poor conception rate in the farmers herd
11. High mortality of the male calves
12. Lack of transfer of technology to the field

Methodology for Estimating Financial Aspect of Progeny Testing

Estimation of selection efficiency

Gross efficiency of selection may be obtained as the ratio of R/C where 'R' is the total gross monetary value and 'C' is the testing cost of male selection programme.

Estimation of gross monetary value

Soller *et al.* (1966) suggested that the long term monetary value of selection responses can be estimated as below:

Monetary return (R) = Genetic response (ΔG) X Monetary value of response X Discount factor X Number of lactations in which it is expressed.

Genetic response to selection

It is estimated by the formula which is as follows:

$$\Delta G = 0.5\, i\, \sigma_g\, r_{IA}\, (1\text{-}P)$$

Where,

i is selection intensity

σ_g is the genetic standard deviation of milk production

r_{IA} is the accuracy of selection

(1-P) is the proportion of population mated to proven sires.

Only the two paths from sire to sons and sire to daughters have been considered and hence the response is multiplied by 0.5

Monetary value of the response

The monetary value of the response is the income from the increased milk yield after deducting the cost of additional feed. The number of lactations (M) in which the genetic response is expressed is calculated as

M = (Population size) (1-P) (Replacement rate) (Average herd life)

Discount factor

The discount factor is necessary because the benefits of selection are obtained several years after the initial investment of resources. The value of responses is discounted by 10% per annum, according to the formula

$$\text{Discount factor} = (1/\ (1+d))^n$$

Where, 'd' is the discount rate and 'n' is the number of years from the commencement of testing to the time in which the response is obtained.

Cost of progeny testing

The technique of artificial insemination with frozen semen has greatly increased the efficiency and utility of progeny testing. A study of the cost structure of this system is necessary to develop a suitable testing programme. VanVleck (1964) and Hinks (1970) studied the financial aspects of progeny testing of bulls.

The variable cost of testing a male for the purpose of progeny testing is composed of the following (Mahla *et al.*, 1981);

$$\text{Total cost} = B + M + P + S + DP + MR$$

Where,

B = the average price of a bull

M = the maintenance cost

P = the cost of collection and processing of frozen semen

S = the cost of storing frozen semen

DP = the cost of data processing

MR = the cost of milk recording a daughter

The costs reported were as below:

Purchase of a bull, Rs. 2285.00 per bull; Maintenance of a bull, Rs. 8.15 per bull per day; semen collection, 3.49 paise per dose; semen processing, 28.22 paise per dose; semen storage, 21.69 paise per dose of frozen semen per annum; data processing, Rs. 2000 per bull and milk recording, Rs. 5 per animal. Handique (1976) calculated similar costs for bull maintenance, dilute semen collection and processing at Central Semen Bank, Khanapara, Assam. Feed accounted for about 46% of the total maintenance cost. Fixed cost accounted for about 80% of cost semen processing and collection. Liquid nitrogen accounted for 55.23% of the cost of storing frozen semen.

The Table (8) presents the total cost of testing eight sires in population sizes of 400, 1600 and 5000 breedable females.

The costs of bull purchase and maintenance rapidly decreased from 66 to 17% of the total cost as the population size increased.

However, the most striking feature was the increase in cost of storing frozen semen from 4 to 52% of the total cost with increase in population size. The increase in cost of processing semen was moderate. In contrast the genetic gain from one generation of sire selection increased from 1.48% of mean (1563 kg) in the small farm population to 5.91% in medium size population of 1600 breedable females. This gain was mostly due to increase in the proportion of population mated to proven sire (1-p) from 0.2 to 0.8. Genetic gain in the large population of 5000 was not proportionately so high as in the medium size population because the only factor increased was accuracy of selection (r_{IG}). The cost of testing per kilogram of increase in total milk production was highest (Rs. 98.10) in the smallest population. It was drastically reduced to Rs.2.80 in the medium size population of 1600 breedable animals. In the largest size population of 5000, the cost was only Rs. 1.44 for every kilogram of increase in total milk production which was quite economic. However, the reduction was not that large because of high cost of storing frozen semen in the case of large population. It is thus apparent that progeny testing in a single herd or small population is a costly proposition and may even fall short of the monetary return expected from the improvement in milk production.

Basu and Mahla (1980) analysed economically the three ways of increasing the genetic progress. They were:

1. Enlarging the size of the population
2. Testing more bulls
3. Increasing utilization of proven sires.

In method 1 it was found that with the increase in population size, the testing cost increased at a diminishing rate but the monetary return increased at a multiplicative rate (Table 8). This showed that the gross efficiency of selection increased with increase in size of population.

In method 2, the cost of testing increased linearly as the number of bulls tested increased. The most interesting feature of the results (Table 8) is the contrast between the diminishing rates of increase in costs from one alternative to another. The lack of response to intense selection could be attributed to:

a) A loss of testing accuracy with reduction in the size of the progeny groups, and

b) Because a non-linear increase in selection rate, viz., a change in the rate of selection from 1 in 3 to 1 in 4 represented a greater increase in the selection pressure than a change from 1 in 4 to 1 in 5.

Table 8: Cost of testing a bull and genetic gain under three population sizes

Particulars	Population size		
	400	1600	5000
Cost of bull (Rupees)	2285 (40.37)	2285 (22.11)	2285 (10.41)
Cost of maintenance (Rupees)	1467 (25.92)	1467 (14.20)	1467 (6.69)
Cost of semen processing (Rupees)	279 (4.93)	1116 (10.80)	3488 (15.90)
Cost of semen storage (Rupees)	229 (4.05)	3665 (35.47)	11452 (52.19)
Cost of data processing (Rupees)	1000 (17.67)	1400 (13.55)	2000 (9.11)
Cost of milk recording (Rupees)	499 (7.07)	400 (3.87)	1250 (5.70)
Total per bull	5660 (100)	10333 (100)	21942 (100)
Total cost for 8 bulls (Rupees in thousands)	45.3	82.7	175.5
Selection rate	1/8	1/8	1/8
Proportion of population test mated	0.80	0.20	0.20
Progeny group size	10	10	31
Heritability of milk yield	0.25	0.25	0.25
σ_P	450	450	450
Mean milk yield (Kg)	1563	1563	1563
ΔG (% mean) / generation	1.48	5.91	7.78
Cost per Kg milk increase (Rupees)	98.10	2.80	1.44

Figures in parenthesis indicate percentages of the total cost

The artificial insemination provides an opportunity to use proven sires extensively. Method 3 made use of extra capacity of proven sires in producing semen to increase selection intensity by requiring less number of bulls. Improved sire utilization increased the rate of sire selection from 1 in 4 to 1 in 8 (Table 9 & 10). This increased the cost of testing as less number of bulls selected meant more number of doses of semen per bull was to be stored for future use. Improved bull utilization increased genetic progress per year, which was mostly due to greater selection intensity. Though the discounted monetary, return increased with improved bull utilization, the benefits were nullified by the increase in cost of testing, particularly storing of frozen semen. The gross efficiency of selection of the three alternatives was almost the same. Hinks (1970) also reported that there was little advantage to be gained form adoption of semen storage programmes in relation to the benefits to be obtained from improved bull utilization. In contrast, if all the bulls were 'laid off' till their evaluation as is the policy in some farms, the cost of testing though costlier would be the same as with frozen semen. Increase in selection pressure and consequent increase in monetary return by wider utilization of bulls was nullified by increased cost.

Table 9: Effect of increasing male selection intensity on cost and efficiency of selection in a breedable population of 10000 buffaloes (30% mated to young sires)

Particulars	Alternatives					
	1	2	3	4	5	6
No. of bulls tested	12	16	20	24	28	32
No. of bulls selected	4	4	4	4	4	4
Proportion selected	1/3	1/4	1/5	1/6	1/7	1/8
Progeny group size	50	38	30	25	21	19
Genetic progress/ year (% of mean)	0.61	0.73	0.78	0.80	0.82	0.84
Cost of testing (Rs. in thousands)	114.6	151.2	187.7	224.3	260.8	297.4
Discounted return (Rs. in thousands)	316.69	377.71	399.44	410.88	417.81	427.27
Gross efficiency (R/C)	2.76	2.50	2.13	1.83	1.60	1.44
Relative efficiency (Δ R/ Δ C)	–	1.67	0.69	0.31	0.19	0.26

Sharma (1985) also studied the economics of three different methods for increasing the selection intensity among males for maximizing genetic gains for Tharparkar cattle

1. Increasing the selection intensity among young bulls and increasing the testing capacity of the population keeping progeny size constant
2. Increasing the selection intensity among young bulls and decreasing the progeny group size, keeping the testing capacity as constant
3. Increasing the usage of proven bulls and the results showed that

Table 10: Effect of increased bull utilization on the cost and efficiency of breeding programmes (20% of the population mated to young sires)

Particulars	Alternatives		
	1	2	3
No. of bulls tested	16	16	16
No. of bulls selected	4	3	2
Selection intensity	1.271	1.400	1.628
No. of doses of frozen semen stored	65,280	87,018	130,560
Genetic progress/ year (% of mean)	0.75	0.82	0.96
Cost of testing (Rs. in thousands)	154.7	166.2	189.3
Discounted return (Rs. in thousands)	474.67	515.16	590.11
Gross efficiency (R/C)	3.07	3.10	3.12
Relative efficiency (ΔR/ΔC)	--	3.52	3.24

In a small herd (300 cow population), the progeny testing is not an economic proposition. The small herds can be a part of the associated herd progeny testing programme, wherein number of herds formed into a grid for common objective of testing certain young sires.

Breeding programme leading to the maximum profit is a sub-optimal programme than for maximum genetic progress. The cost of progeny testing is directly proportional to the number of young bulls taken in the programme. It is more profitable to test less number of bulls with more daughters and selection rate 1:4 among sires.

Set Wise Progeny Testing Programme Layout

Progeny Testing Programme layout is designed here will have the following assumptions

1. Single organized herd (herd size = 500)
   herd size includes breedable females, breedable males, calves and heifers etc.
2. Breedable female population = 300
3. Bull : cow ratio = 1:30
4. Conception rate: For buffalo = 45%
   - For indigenous cattle = 55%
   - For crossbred cattle = 65%
5. Mortality = 10%
6. Abnormal calving = 2%
7. Sex ratio (Male: Female) = 1:1
8. Test cycle (For each set) = 18 months (Buffalo)
   24 months (Cattle)
   30 months (Crossbreds)
9. Daughters reaching Age of First Calving = 2/3rd
10. Frozen semen doses = 5000 (Herd P.T.) + 3000 (Field P.T.)

Set Wise Progeny Testing Programme Layout

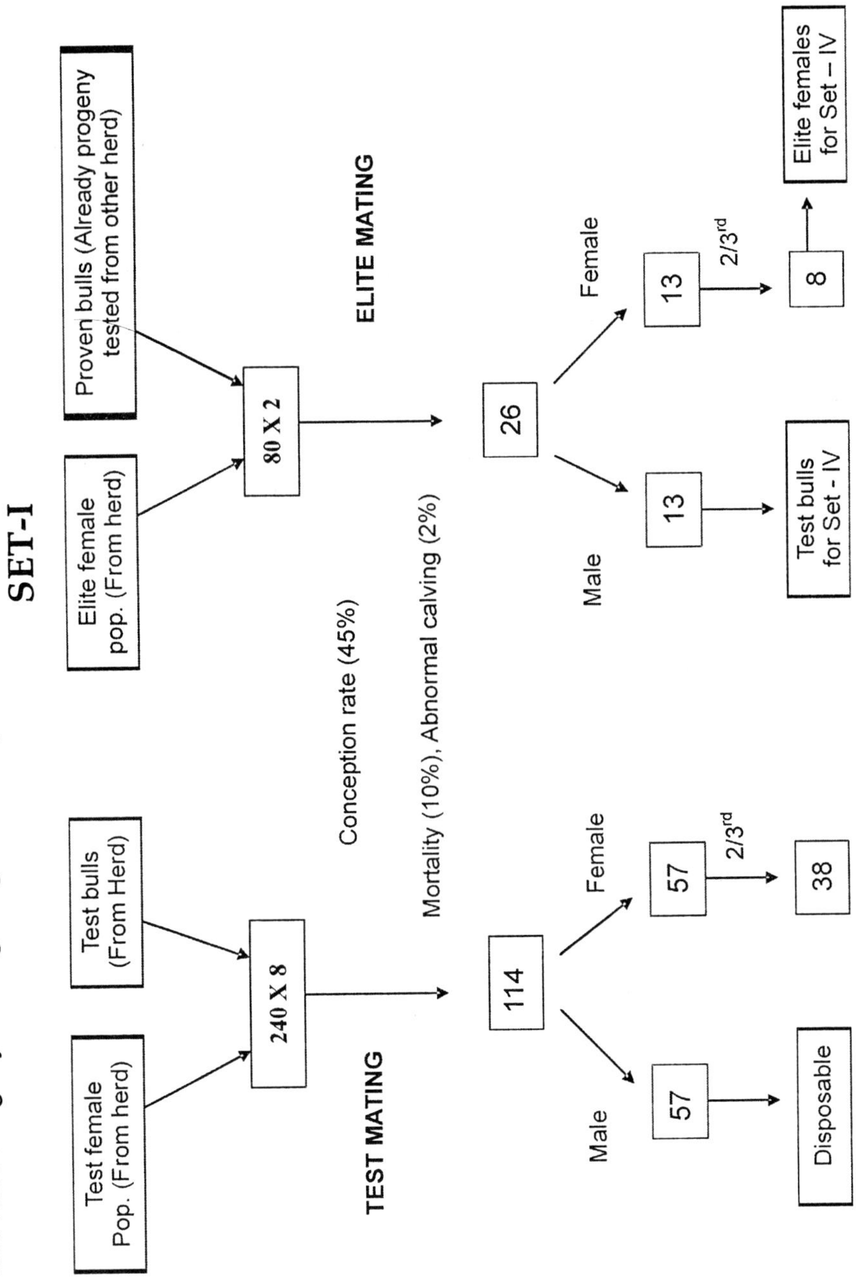

Set Wise Progeny Testing Programme Layout

SET-II

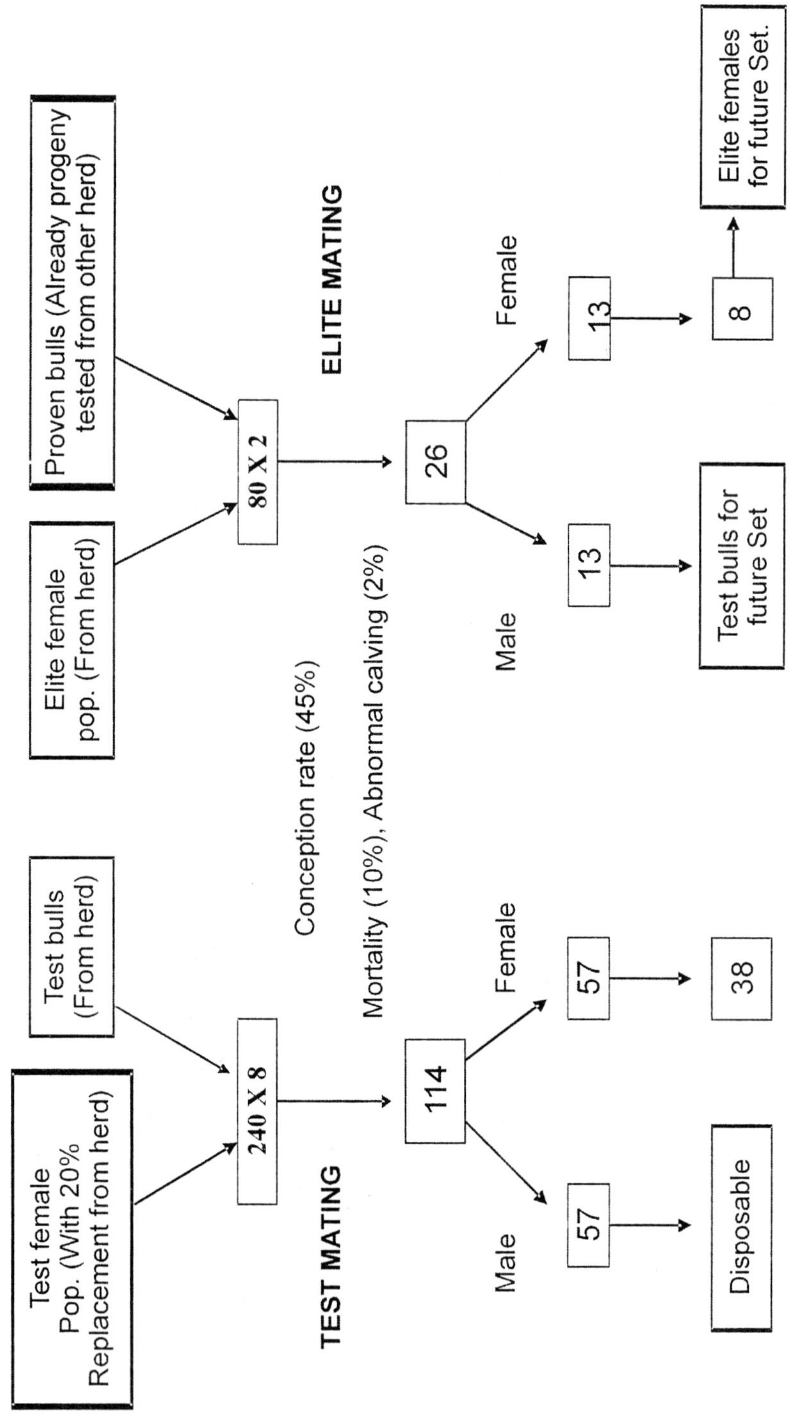

Set Wise Progeny Testing Programme Layout

SET-III

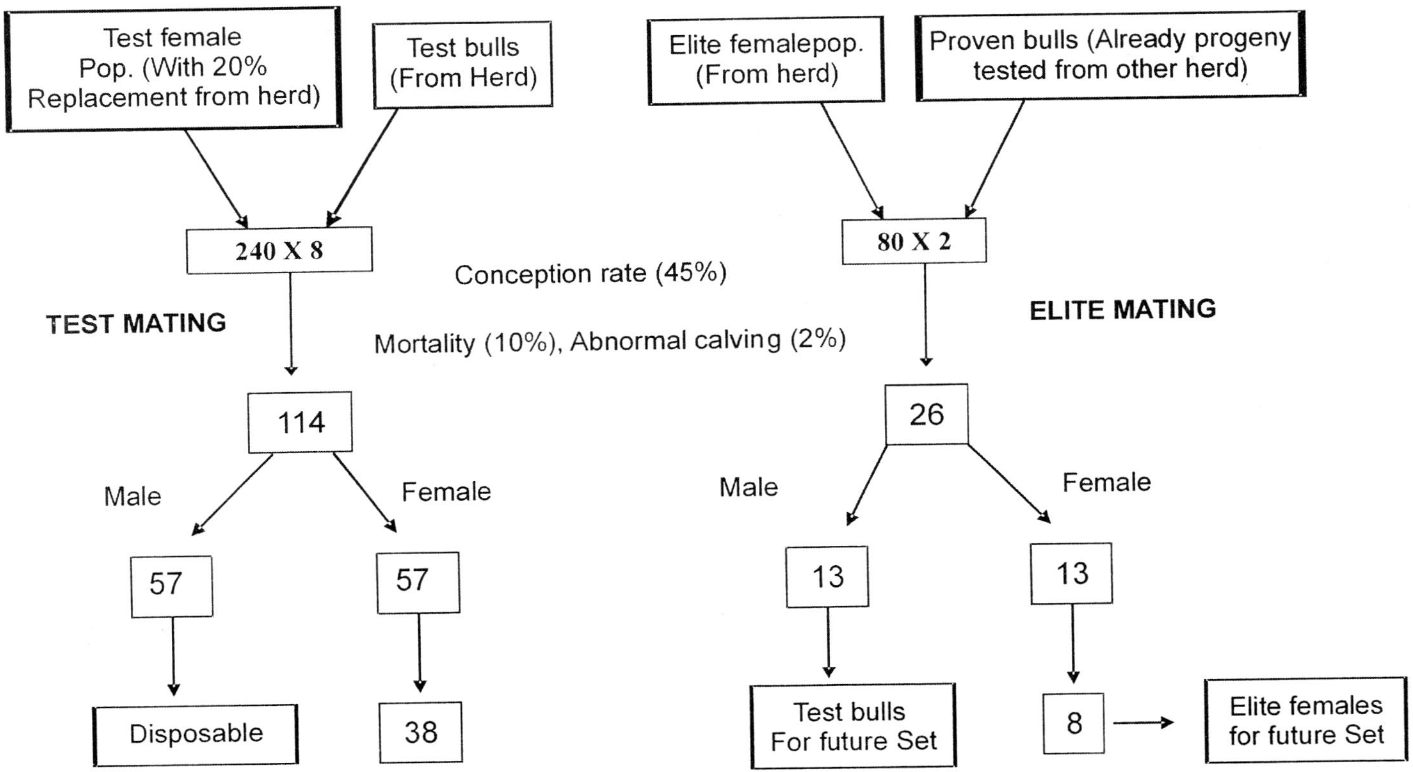

Set Wise Progeny Testing Programme Layout

SET-IV

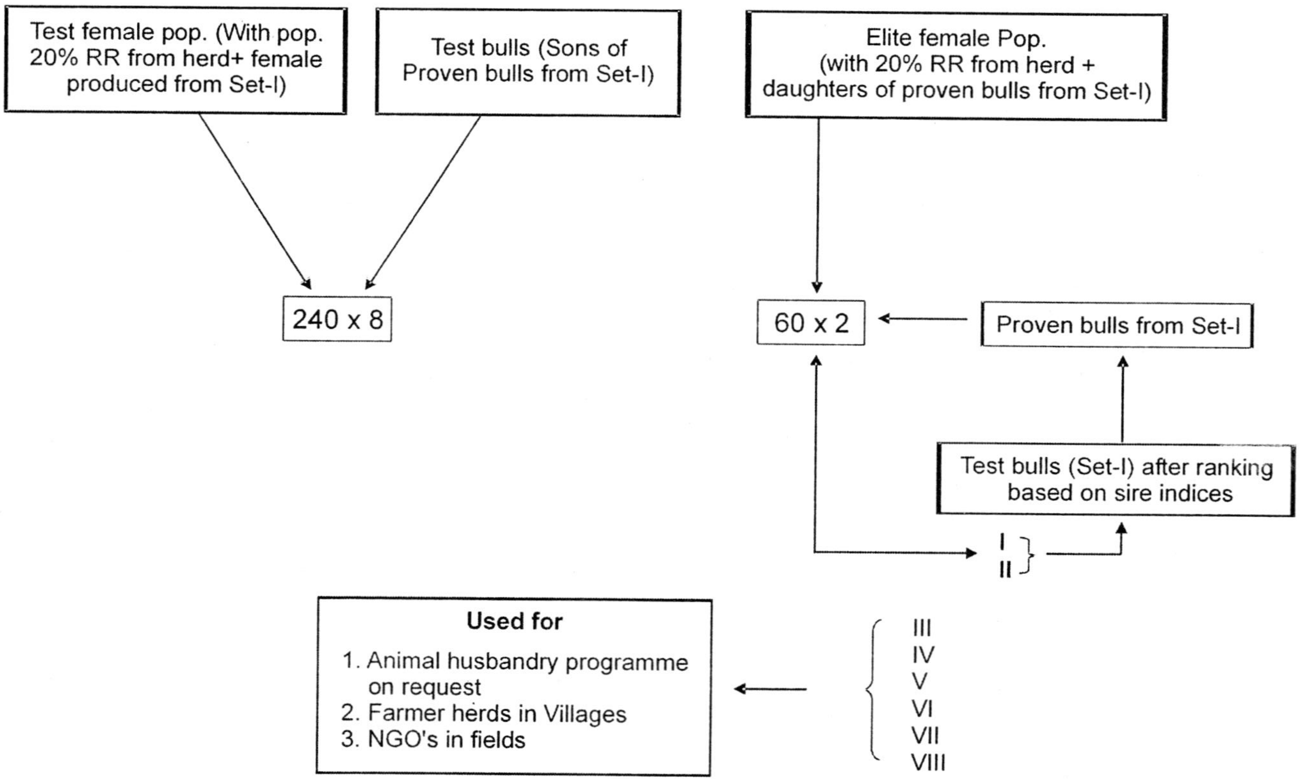

Summary

Various scientific programmes, for genetic improvement and to improve the productivity of the native breeds, were initiated and a number of farms of these breeds were established for production of superior quality bulls. Bulls were selected on the basis of breed characteristics, body conformation and milk yield of their dams wherever available and it was recommended that infrastructure should be developed to test the bulls on the basis of their daughters' performance before they are used in developmental programmes. Accordingly, progeny testing programmes for some of the improved indigenous breeds, like Tharparkar and Red Sindhi under Centrally sponsored schemes and Sahiwal, Hariana and Gir, under the State sponsored schemes were initiated. Some bottlenecks exist in applying/ implementing progeny testing mainly because of the small herd size used. Non-existence of deep-freezing facilities at these farms also contributed to the failure of these schemes because by the time bulls became available after test, they were too old to donate any semen.

The study on the financial consequences of the progeny-testing programme reveals that there was a little advantage to be gained from adoption of semen storage programmes in relation to the benefits to be obtained from improved bull utilization (Hinks, 1970). In contrast, if all the bulls are "laid off" till their evaluation as is the policy for some farms, the cost of testing though costlier would be the same for all alternatives. Increase in selection pressure by wider utilization and consequent increase in monetary return is obtained without increasing cost. A large population should be involved in progeny - testing sires so that greater part of population is available for breeding with proven sires and the cost of progeny testing programme is directly proportional to the number of young bulls progeny tested. It is more profitable to test less number of bulls with more number of daughters. Further, if semen is to be stored, improved bull utilization does not extend the limits of profitable selection.

□□□

Chapter 6

Breeding Strategies and Performance Recording Under Field Conditions

Field Progeny Testing

Sustained genetic improvement can be achieved only through continuous use of bulls of high genetic merit evaluated through progeny testing. The genetic improvement programme undertaken on single herd basis with limited population size could not provide sufficient number of progeny tested bulls with higher accuracy and intensity of selection. Networking of various cattle and buffalo breeding centre maintaining the same breed (multi herd progeny testing) though improved accuracy and intensity to a limited extent but could not provide required number of tested bulls for bringing genetic improvement in a large population of cattle and buffalo. Therefore, there is need to undertake well planned improvement programme to identify large number of superior breeding bull and evaluating their genetic merit. To meet the huge requirement of superior bulls through on-going institutional herds progeny testing programme may not be possible, therefore, large scale progeny testing of bulls under field conditions should be taken up in collaboration with different research and development organizations like State Livestock Development

Board, State A.H. Department, State Milk Producers Co-operatives union, Research Institutes, State Agricultural Universities and Private Livestock Breeders maintaining large herd. Table (11) compares the progeny testing scheme at both organized herd and field progeny testing.

Table 11: Comparison between Organised Herd and Field Progeny Testing

S.No	Herd Progeny Testing	Field Progeny Testing
1	Individual evaluated based on their progenies production performance raised under organised herd level	Individual evaluated based on their progenies production performance raised under field or farmer's herd level
2	Less number of bulls can be tested.	More no. of bulls can be tested
3	Number of progenies per bull is limited.	Number of progenies per bull is more.
4	Progenies are raised in same environment	Progenies are raised in different environment
5	Efficiency & accuracy is less	Efficiency & accuracy is more.

Requirements for Field Progeny Testing Programme

- Select the village where no breeding bull is available for Natural Service
- A.I. centre should be approachable by road
- Distance of A.I. centre should not more than 40-50 Kms
- Sufficient number of A.I. worker need to be appointed, for covering the animals
- Veterinary aid is required to attract farmers for A.I. & to provide health cover to registered animals
- Animal performance should be recorded. Kaura et al. (1983) suggested that fortnightly or monthly recording was most efficient & accurate for predicting milk yield.
- Sufficient incentives to be provided to farmers to keep the heifer born to Progeny Tested bull till first lactation records are available

Field Performance Recording

The performance recording system of farmer's herds under field condition requires all the basic infrastructure facilities a dairy herd improvement or development programme needs. Therefore all the existing or proposed dairy improvement or development programme need to be integrated with performance recording of farmers herds. Thus to start with, all those areas with high density of relatively high producing dairy herd units as well as the pockets where research and development organization's breed improvement and development programmes are going on may be identified for developing and implementing the field recording programme

Steps needed to be taken in recording field data

1. Development of simple and convenient recording formats for collection of data on farmer's herd.
2. Testing and adoption of animal identification methods i.e. tattooing, branding, ear tagging with plastic tags etc.
3. Data on herd performance could be collected at different levels. Milk recorders should collect primary data on socio-economic levels of farmers and their herd status, performance and management status etc.
4. Monitoring and Compilation of information on AI's, pregnancies, calvings, animal identification as well as surprise checks on accurate collection of data.
5. Collection of data from field centres to be transferred to the data storage and processing unit after proper scrutiny, editing and updating.
6. For active participation of farmers in the herd improvement and performance recording programmes, formation of farmers co-operative milk unions, Herd/ Breed Association should be promoted.

Organizational Setup for Sustainable Field Progeny Testing Programme

For implementing breed improvement programme under field conditions we must accept and understand the constraints. To overcome them, we should develop an appropriate network of infrastructure along with field extension services and adopt suitable methodology of performance recording. The basic organizational infrastructures needed for such programmes are as under

1. Central coordinating agency

A national level central coordinating agency should be established to co-ordinate these programmes by working in close association with State Animal Husbandry Departments, State Agricultural Universities, Research Institutes and other participating organizations. It should formulate a detailed technical programme and monitor for its uniform implementation by participating units. There should be central data repository with facilities to compute time to time evaluation for monitoring and coordinating the programme.

2. Coordinating progeny testing units

Different progeny testing units should be established in different regions for planning and execution of technical programme. Information on animal's identification, its age, sire, dam, data on growth and milk production traits etc are to be generated, compiled and stored in computer. These units should have data on cattle owner's socio-economic level and other managemental practices for developing correction factors for non-genetic differences if any for evaluation of the test bulls. This will help in selection of top ranking bulls and identification of elite cows for nominated mating to produce future bulls for progeny testing.

3. Germplasm centres

The promising young bulls selected on the basis of their pedigree performance and procured from bull mother farms or purchased from farmers or other agencies should be reared and further propagated at germplasm centres. There should be sufficient number of male calves initially selected for required numbers of breeding bulls in each batch and should be tested for various communicable diseases and genetic disorders.

4. Networking of extension centres with veterinary, inseminating, milk recording and supervisory staff

To provide the AI facilities at the doorstep of farmers, AI centres should be established at suitable locations. All the animals inseminated should be followed for their pregnancy diagnosis, treatment of reproductive problem animals, calving, sex and birth weight. The milk recording may be carried out by trained milk recorders employed on contract basis or by the animal owner himself. The milk recording schedules may be prepared well in advance and given to the milk recorders for milk recording.

5. Incentives to the farmers

The small dairy herd units, which are important component of our agricultural economy, need to be given all possible technical support and other incentives for their participation in field recording and bull evaluation programmes. It will be desirable to associate relatively large size herd units belonging both to farmers and the research institutions.

6. Involvement of cooperatives and NGO

Different organizations engaged in the dairy development programme should be identified and their participation and cooperation along with pooling of the resources must be ensured in developing and implementing the field recording programs. Committees of the representatives of these agencies may be constituted at the state and national level for proper monitoring and coordination of dairy herd improvement. The farmers identified for a particular herd/breed improvement programmes may be organized to form their cooperatives. The field recording programmes should be closely linked with the dairy herd improvement, development and extension programme. There should also be enhancement of the role of NGOs in performance recording at field level.

❑❑❑

Chapter 7

Breeding Strategies Under Different Animal Production Systems

It is essential to plan strategies in such a manner that available resources in all the agro-ecological zones of the country be exploited judiciously and utilized sustainably for further enhancing the productivity of animals. For developing strategic planning for enhancing overall productivity under different agro-climatic regions of the country, India has been divided into 15 agro-climatic regions based on physiographic and climatic conditions. These zones are further sub-divided into 126 agro-climatic sub-zones by taking into consideration rainfall, temperature, soil-type, topography, existing cropping pattern and farm animal resources for developing policies on regional basis. Mixed farming is practiced in nearly all Agro Ecological Zones.

The principal objectives of the farmers engaged in mixed farming include:

i) Complementary benefit from an optimum mixture of crops and livestock farming

ii) Spreading income and covering risks over both crops (primary) and livestock (secondary) production

iii) Scope to adjust crop and livestock ratio to social and economic needs and opportunities.

Animal Production Systems (APS)

According to FAO, all input-output relationships viz. biological, climatic, economic, social, cultural and political factors, which combine to determine the production of a particular livestock enterprise at a particular location is referred as Animal Production System. APS range from very little animal husbandry to very intensive management systems where livestock keepers control feed, climate, disease and other factors. Animal production system in India can be subdivided mainly into Dairy, Meat, Wool, Draft production apart from skin and hide production and dung production.

Here we shall focus on sustainable breeding strategies for enhancing productivity of different dairy cattle populations under different animal production systems in India. Since these strategies depends upon the APS first let us see how these animal production systems are classified. Broadly, three Production Systems viz. Low input, Medium input and High Input dairy production systems have been classified.

Low-Input (Extensive) Animal Production System

A production system where one or more rate-limiting inputs impose severe pressure on livestock (continuously or variably), resulting in low survival, low reproductive rate or less output. This system is characterized by low input use and low productivity. Farmers who are poor in resources especially small / marginal farmers and landless labourers maintain majority of the livestock under low-input animal production system, across all the states and in different agro-climatic regions. These farmers maintain only 1-2 cattle/ buffaloes under this system. Around 95% of such animals which belongs to locally adapted breeds or may be non-descript populations survives because they are better adapted to the local ecology, use of crop by-products and low requirement in terms of feed, fodder, health care, housing and labour. Milk production and draft animal power are the main production activities. Under this system, most of the output (milk) is consumed by the family and less than 50 percent of the milk is marketed.

Medium-Input (Semi-Intensive) Animal Production System

In this type of production system management of the available resources has the scope to overcome the negative effects of the environment. Although it is common that one or more factors in this system can limit the output, survival and reproduction in a serious manner. The medium dairy farmers having better resources compared to poor farmers, they provide medium levels of input and management under this system. The farmers generally maintain a mix of local cattle/ buffaloes, a few crossbred cattle and some buffaloes belonging to a good dairy breed under this system. Small and medium family farms under this system, market more than 50 percent of the output (milk).

High-Input (Intensive) Animal Production System

All rate-limiting inputs to animal production can be managed in high input animal production system to ensure high levels of animal output, animal survival and their reproduction. Output is constrained primarily by managerial decisions. The farms following intensive production system are using technology and inputs developed locally or imported, and the production of these farms approaches the western standards. Intensive system of animal production for cattle and buffaloes is mainly restricted to State/ Central government livestock farms at State Agriculture Universities (SAUs) and ICAR institutes but some farms run by the NGOs and progressive commercial farmers also uses this system. Under this system, crossbred cattle and buffaloes having higher levels of productivity are maintained with intensive management in terms of breeding, feeding and health care. Large commercial farms or companies maintaining large herds (more than 50 heads) of cattle herds belong to this category and all the milk produced is marketed as liquid milk or milk products.

Breeding Strategies

In order to develop the sustainable breeding strategies for conservation and improving the productivity it is imperative to have basic and comprehensive information. This includes analysis of existing activities, farmer/breeder's perceptions and socio-economic levels, livestock production systems, available animal genetic resources (type /breed/genetic group of animal population in a given agro-climatic region) and their production potentials, available breeding organization, available infrastructure and development facilities. Based upon the collected and collated information of the area/region as well as the types of animal population following sustainable animal breeding technologies can be adopted for conservation and genetic improvement of bovine genetic resources under different animal production systems.

1. Crossbred Cattle Under Different APS

As a result of crossing non-descript cattle or local breeds with high yielding exotic cattle breeds is the quickest and most effective means of achieving high yielding animals. A large number of half-bred and higher crosses of crossbred animals have been produced at organized farms and under rural households in different agro-climatic zones of the country. The crossbred animals are reared mainly under semi-intensive and intensive animal production systems. Accordingly, the breeding policy for existing crossbred animals under semi-intensive and intensive animal production systems is suggested as follows:

a. Breeding policy for crossbred animals after F_1 generation under semi-intensive APS

Under this production system, it is advisable to restrict exotic inheritance level between 50 and 62.5%. Therefore, the F_1 crossbred females under semi-intensive production system can be bred with the semen of genetically superior progeny-tested crossbred males having exotic inheritance between 50 and 75% produced through inter-se mating among crossbred animals. Holstein Friesian is the breed of choice in the irrigated plains. Jersey is the breed of choice in hilly terrain and coastal areas for crossbreeding.

b. Breeding policy for crossbred animals after F_1 generation under intensive (High Input) APS

High input production system is mostly available in milk shed areas around peri-urban and industrial towns where good market for milk and milk products is available and adequate quantity of quality feed and fodder resources exist. Under intensive production system, higher levels of exotic inheritance between 62.5 and 75% can be sustained. The F_1 females in the field can be bred with the semen of genetically superior progeny-tested males of exotic breeds with high progeny test index to produce progeny with 75% exotic inheritance. Further, the progeny tested crossbred males having exotic inheritance between 50 and 75% produced through inter-se mating can be used to sustain the exotic level between 62.5 and 75%. The semen of exotic bulls of high transmitting ability for milk yield and milk constituents should be used. The crossbreeding of non-descript zebu cows with semen of exotic dairy cattle breeds will result in enhancing milk production by 5 to 8 times to that of non-descript cows, reducing age at first calving and shortening calving intervals in crossbred progenies. Holstein Friesian is the breed of choice in the irrigated plains. Jersey is the breed of choice in Hilly terrain and coastal areas for crossbreeding.

2. Nondescript Cattle under Low-Input APS

The local nondescript cattle are reared mainly under Low-Input Production System across the different agro-climatic zones where quality feed and fodder resources are not available in sufficient quantity. The farmers are not resource-rich. Infrastructure facilities are inadequate. Marketing facilities for sale of milk and milk products are poor. The nondescript cattle can be genetically improved by grading up using elite bull of well known indigenous breeds like Sahiwal, Gir and Red Sindhi for milk production. Through continuous grading up nondescript stock will be transformed in to well defined breed after seven generations but maximum genetic gain is received during first three generations after which the gain is relatively slow.

3. Well-Defined Indigenous Dairy Cattle Breeds under Medium to High Input APS

The animals relatively with high producing ability belonging to well-defined indigenous dairy and dual purpose cattle breeds are maintained under intensive production system at organized farms and mainly under semi-intensive management system in farmers' herds. These breeds can be genetically improved through selective breeding. High-pedigreed young bulls with high Expected Predicted Difference (EPD) and genetically proven bulls with high Breeding Value (BV) should be utilized for the purpose in the breeding tracts of different zebu cattle breeds. For example, Gir and Kankrej for Gujarat; Rathi, Nagori and Tharparkar for Rajasthan; Hariana for Haryana, parts of Punjab, Western U.P. and Rajasthan and Ongole for Andhra Pradesh may be utilized.

4. Well-Defined Buffalo Breeds under Medium to High Input APS

The relatively high yielding buffaloes of well-defined buffalo breeds are maintained under intensive production system at organized farms and mainly under semi-intensive management system in farmers' herds in the breeding tracts of different buffalo breeds. Selective breeding is practiced for bringing about genetic improvement in buffalo breeds. The genetic improvement in indigenous buffalo breeds for higher milk production, reduction in age at maturity, reduction in service period, dry period and calving interval will lead to higher economic returns to the farmers. By selective breeding, through networking of multi-herds of a particular breed, it is expected that genetic improvement in milk production will be achieved.

5. Local / Nondescript Buffalo Breeds under Low to Medium Input APS

The low producing, local non-descript buffaloes are generally reared under low to medium input Production System in areas where feed and fodder resources and milk and animal marketing facilities are moderately available. The production potential of low producing non-descript buffaloes can be increased rapidly through mating with superior sires of improved breeds like Murrah, Surti and Mehsana. Murrah is recommended for grading up of medium body sized non-descript buffaloes. This programme is expected to increase the milk production of village non-descript buffaloes by 2 to 3 times in early generations of grading up. The grading up of non-descript buffaloes yielding on an average 500 kg with improved buffalo bulls having genetic potential of 2000 kg or more will yield to an average of 1250 kg in first generation. Thus, through grading up with superior breeds in four to five generations, the low producing non-descript buffaloes can be replaced with relatively high producing buffaloes conforming to the characteristics of well-defined breeds.

Methodologies for Performance Evaluation & Selection of Animals

Methodologies developed for selection of adult females based upon their Expected Producing Ability (EPA) as suggested by J.L. Lush, 1945 for ranking of adult cows. Young males based on their Expected Predicted Difference (EPD) using pedigree information, physical attributes and subsequently evaluating them on the basis of their progeny performance etc. could be utilized for improvement of cattle and buffalo herds maintained on organized farms of Central/ State Government and other developmental agencies. There are two most popular methods of estimating the EPD one is paternal dam method and second is paternal dam and grand dam method. For bringing effective genetic improvement, rigorous selection pressure must be ensured on the basis of milk production of cows. Here it is very important to increase the herd replacement rate and minimize involuntary culling on the basis of reasons other than milk production (like mastitis, lame cows, premature deaths etc.) For this, proper herd management practices such as suitable housing, proper adoption of reproductive health management practices, adequate availability of quality feed and fodder, timely disease control measures to minimize the mortality need to be employed. Since large proportion of genetic gain is resulted through proper selection of superior breeding bulls, it should be ensured that the young bulls are the progenies of elite mating and subsequently be progeny tested on large test population associating multiple organized herds or farmer herds. Elite cows for nominated mating may be chosen from organized herds and from farmers' herds. For implementing the strategies in undertaking large scale genetic improvement programmes on widely distributed population of cattle and buffaloes in the country, we require huge number of genetically superior breeding bulls along with adequate networking of infrastructure. According to an estimate, to cover even 30% of existing breedable bovine population in the country, for breeding through AI as many as 1050 proven bulls of crossbred cattle, 5700 proven bulls of well-defined indigenous cattle breeds and 11400 proven buffalo breeding bulls are required. Therefore, for accomplishing the huge task of bringing about genetic improvement of cattle genetic resources in the country through above proposed breeding methodologies besides developing basic infrastructure facilities, identification of important breeding services and animal developmental organizations, the following action plan for proper implementing the programmes is suggested

1. Identification and establishment of breeding services organization (Centre/ State level) for overall monitoring and implementation of breeding programmes.
2. Identification of operational areas/ regions for the breed/genetic group/animal population to be improved upon.
3. Infrastructure development in the operational areas including establishment of AI centres, animal health centres, milk procurement and extension centres fully equipped with trained man power i.e. veterinary, para-veterinary staff so as to enlarge network of breeding facilities for covering large number of breedable bovine animals.

4. Establishment and strengthening of breed nucleus herds/bull mother farms, young bull rearing centres, semen collection and cryo-storage banks.
5. Establishment of data bank to carry out activities on animal identification and performance recording linked with progeny testing programme for selection of large number of high genetic merit bulls.
6. Networking of breed-specific organized government/ private cattle and buffalo farms and large progressive farmer's herds for testing of large number of breeding bulls and linking with performance recording and progeny testing of bulls under farmer herds in village conditions.
7. For faster multiplication of superior germplasm and large-scale dissemination, selected organized nucleus breeding herds should adopt open nucleus breeding system with or without MOET and other emerging reproductive and genetic techniques.
8. Formation of the milk producers' co-operative unions/breed societies/NGOs for participation of farmers/ breeders in the activities on breed development and improvement as their own programme.

□□□

Chapter 8

Present Status, Success/Failure and Future Road Map for Crossbreeding in India

Efforts for Crossbreeding

Cross breeding in India started as early as in 1875 by European missionaries to develop the "Tylore breed" in Patna by using short horn breed on native cows. In early 1900 the project on crossbreeding in Nilgiri district of Madras was started by European planters who brought Ayrshire, Friesian and Jersey. Organized work on crossbreeding started in 1919 at Honsur cattle farm by Madras government with an objective of evolving a new dairy breed by crossing Aryshire and Red sindhi cows. In 1927, Forman at Military Dairy Farm reported that HF crosses are more suitable and adaptable than Ayrshire crosses. During 1924-34, Naini Agricultural Institute at Allahabad started the cross breeding programme with four exotic breeds Holstein, Brown Swiss, Guernsey and Jersey and five indigenous dairy breeds Kankrej, Gir, Sahiwal, Haryana and Red Sindhi were studied. Kartha (1954) reported that HF crosses are superior to both Ayrshire and short horn.

In 1955, the crossbreeding work was started in Panganur cows with Kerry bulls at Govt. Livestock Farm, Palamner (Andhra Pradesh). In 1961 the Indo Danish project started in Hesserghata (Karnataka) and highest milk production and shortest calving intervals were observed in Red Dane crosses. In 1963 (Kerala) and subsequently in 1971 (Punjab) Indo Swiss project was started and Brown Swiss (BS) was introduced in India, and crosses form BS were having less age at first calving and lower calving interval. During 1980-82 Karan Swiss and Karan Fries breeds (Figure 5) were developed by National Dairy Research Institute, Karnal Haryana. Karan Swiss is a cross of Sahiwal with Brown Swiss while, Karan Fries is a cross between Tharparkar and HF. Karan Fries crossbred cows were preferred by farmers over Karan Swiss crossbred cows.

Jersind (Jersey and Red Sindhi) was developed at Naini with 3/8 to 5/8 exotic inheritance levels.

Military Dairy Farm in 1985 developed the Frieswal (HF X Sahiwal) with 3/8 to 5/8 level of exotic inheritance.

Sunandini was developed in Kerala during the Indo Swiss project by crossing the local cattle with Brown Swiss.

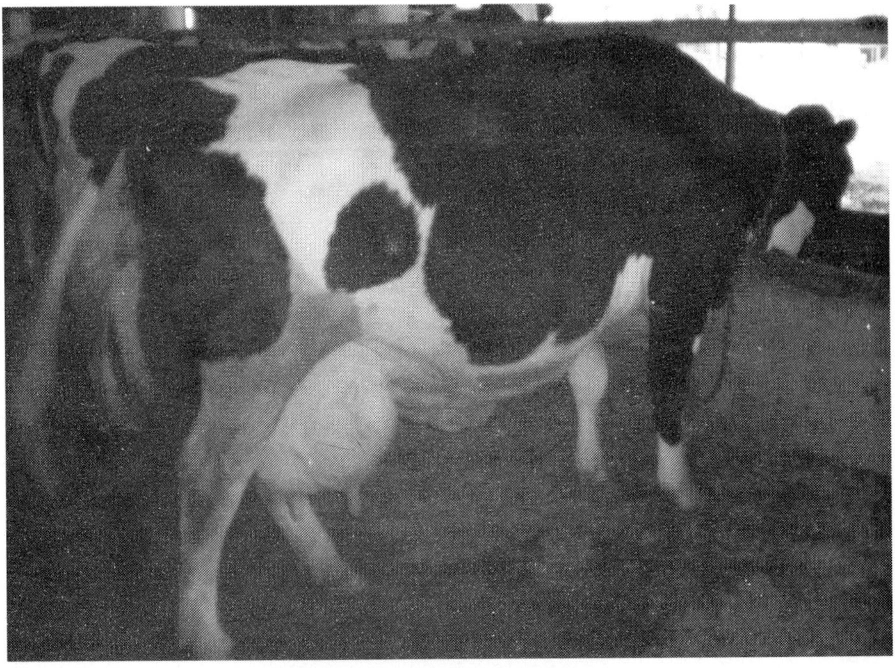

Figure 5: a) Karan Fries Cow

Figure 5: b) Frieswal

c) Jersey Crossbred

Figure 5: Major Synthetic Cattle Breeds Developed in the Country via Crossbreeding

Present Status of Crossbreeding

Choice of germplasm for crossbreeding

Germplasm of Jersey and Holstein-Friesian will continue to be two germplasm of choice to be used for crossbreeding program in cattle. The HF crosses in plains where plenty of feed and fodder is available and Jersey in Hilly and Coastal regions where the small body size of these animals gives edge over HF. Though the crosses of brown swiss also provides a good compromise between milk yield and fat percent and are seen equally successful but farmers choice is not for Brown Swiss (as in the case of Karan Swiss in Northern Parts of India). It has been seen that two breed exotic crosses are more adaptable in our environmental conditions than any triple-cross-breed, keeping in view technical difficulties in breed stabilization, achieving the optimum milk yield and in maintaining the exotic level of inheritance.

In choosing the superior indigenous animals to be crossed with exotic germplasm it is absolutely imperative to improve the productive capacity of indigenous animals with an aim of a healthy calf each year through crossbreeding. This is possible only by increasing the reproductive efficiency of the superior animals. Successful crossbreeding between exotic and nondescript animals encompasses the ability to mate, the capacity to conceive and to nourish the embryo and deliver the viable young ones at the end of a normal gestation period. In fact, interruption in this chain of events leads to failure of the cow either to conceive or the embryo to die or to have a premature delivery of the foetus. The breeding efficiency is a complex phenomenon controlled by both genetic and non-genetic factors, the non- genetic factors being climate, nutrition, and level of management. Therefore, it is necessary to consider the genotype environment interaction (refer Chapter 11 for details) while formulating a breeding programme to increase the overall adaptability of the cross breeds also the animal should be free from all types of diseases.

Crossbreeding policy for cattle

Breeding Policy for cattle must be formulated keeping in view the socio-economic conditions of the farmers. Breeding policy must be aimed at increased productivity of cows by genetic improvement. However, in Indian Scenario it should also aim at conservation of native breeds and to ensure good quality of bullocks for animal draught power in rural areas. In order to achieve self sufficiency in milk production it should be aimed at achieving a higher level of 60%

genetic improvement from present level of 28% by end of year 2015, instead of the year 2010 as envisaged under National Project for Cattle and Buffalo Breeding (NPCBB, for details refer glossary) and to further improve it to 80% by the end of year 2025. Above mentioned targets are expected to be achieved through the consolidated and collective efforts of all the agencies engaged in cattle breeding activity viz. State Animal Husbandry Department, Co-operative Milk Unions, NGOs, Private Sector Agencies and unorganized Artificial Insemination (AI) workers in the various states.

Strategy and Approach

1. For milk purpose, non-descript cattle are crossed with exotic breeds of Holstein-Friesian (HF) and Jersey (purebred and crossbred HF and Jersey bulls).
2. For draft and dual purpose (milk/draft), nondescript cattle are improved with recognized indigenous breeds of the area, e.g. Hariana cattle in Haryana State.
3. For crossbreeding, HF is the major exotic breed, while Jersey is used in less resource full areas with comparatively smaller size nondescript cattle.
4. The level of exotic inheritance in crossbreds shall be 50%. Therefore, female crossbred offspring produced through crossing of Non Descript with exotic breeds are to be mated with crossbred bulls of 50% exotic inheritance.
5. Scientific genetic selection programmes should be adopted and installed to procure breeding bulls of high genetic value.
6. Crossbred bulls should be obtained from bull mother farms through nominated mating and from areas in the field with field performance programmes. In case of shortage, crossbred bulls shall be procured from outside the State.
7. Sahiwal and Hariana breed are used as the indigenous partners to produce the first generation (F_1) of crossbred breeding bulls. Afterwards, crossbred bulls are produced through inter-se mating of the crossbred population is adopted.
8. Progeny testing programmes are needed to be started for the selection of breeding bulls of all the genetic groups (cattle and buffaloes).

9. Bulls of accepted quality, but not required for semen production, are to be disseminated for Natural Services in areas with no or low AI coverage.

Constraints in the Road of Successful Crossbreeding

In order to have the success in crossbreeding programme we have to overcome some constraints like shortage in the availability of feed and fodder (dry, green and concentrate) which is increasing due to the pressure of meeting the demands for human consumption. Farmers are not growing these fodder crops and majority of the animals have to sustain on by-products and left outs from the fields. There is an urgent need to bridge this gap in production and requirement of fodders.

Problems of Poor Farmers i.e. the poor quality animals, poor reach and quality of services, poor health status and spread of diseases, Lack of technical guidance, Weak marketing network and inadequate credit facilities.

Constraints in Livestock Development

1. Non-availability of superior quality breeding bulls.
2. Poor quality of semen produced by many laboratories.
3. Poor skills of paravets - low conception and infertility.
4. Absence of conservation strategy for indigenous breeds.
5. Lack of coordination among different agencies engaged in animal improvement.
6. Poor extension services to motivate small farmers.

Outcomes of Crossbreeding in India

1. A significant reduction in age at first calving and decreased calving interval were observed in the crossbreds.
2. Milk production was increased in first lactation (F_1 crosses).
3. Crosses made from Tharparkar, Sahiwal, Red Sindhi and Gir showed better results than other native breeds under similar tropical conditions.
4. Holstein Crosses were better in Milk production than other temperate breeds.

5. Jersey crosses showed improved reproductive efficiency compared to other exotic breeds.
6. Exotic inheritance of 50% is most ideal.
7. Hetrosis in production during F_1 generation and thereafter Decline in milk production and reproductive efficiency were observed in F_2 and F_3 generation.
8. Two breed crosses were overall better than three breed crosses.

Schemes under Ministry of Agriculture, Government of India

1. Central Breeding Farms: Germplasm Banks
2. Semen Production Laboratories
3. Herd Registration Scheme: Breed conservation
4. Fodder seed production, fodder banks
5. Biological Units: DI Laboratories, Vaccination for control of RP and FMD
6. Dairy Development: Assistance to Cooperatives, Clean milk production, Venture capital

Future Road Map for Crossbreeding

There is urgent need for favourable National Policy comprising

1. Improvement in efficiency and quality of milk and milk products to compete in global market
2. Enable resource poor farmers to take active part
3. Emphasis on Ecological Sustainability.
4. Extension of dairy development to Non-Operation Flood Areas
5. Complete Animal Health Cover - Disease free zones
6. More Credit facilities to livestock keepers.

Points for Successful Crossbreeding

1. Conduct a survey in the breeding tract of the milch breeds like (Gir, Sahiwal, Redsindhi, Tharparkar, Rathi, Nagori) to know the present status, population, herd's, demographic and geographic distribution of breeds, habits and habitant of farmer's, their needs and management practices, availability of land, fodder etc.

2. Appropriate breeding strategies for the genetic improvement of threatened breed of cattle Selective breeding using latest biotechnologies such as Marker Assisted Selection, and Embryo Transfer Technologies for identification of superior breeding animals and their faster multiplication.
3. More emphasis should be given on indigenous breeds of cattle rather than the cross breeding the superior indigenous germplasm in their breeding tracts. Availability of good quality breeding bulls and artificial insemination services to the cattle owner helps in conservation programme.
4. Improve the nutritive value of poor quality roughages and educate the farmer about balance complete feed e.g. concentrate feed block, bricks, etc., using non conventional feed and concentrate in balance ratio.
5. Transfer of technologies to propagate methods of composting, vermi culture, from cattle dung and urine based pesticides, insecticides and medicine etc. Forming a co-operative society under state Govt. supervision to develop a marketing of cattle by-products.
6. Providing suitable subsidy on balance complete feed at time of scarcity and drought periods to save animals from starvation deaths/distress.
7. Above all, the same crossbreeding and upgrading programme may not be possible in every prevailing livestock production system of every state. Regional and location specific features, such as climatic conditions, feed and fodder availability, livestock management practices and market facilities for livestock produce, have to be taken into account, while proposing crossbreeding programme.

□□□

Chapter 9

Advances in Breeding Methodologies for Sustainable Performance

Introduction

India is predominantly an agrarian society where animal husbandry forms the backbone of the national economy. Buffalo and cattle in India are mainly kept for milk production, motive power for various rural farm operations, transport, irrigation and production of manure. The rearing of dairy animals also provides employment and supplementary income to the vast majority of rural households, the majority of whom are landless and marginal farmers.

Comparative performance evaluations of various cattle and buffalo breeds have clearly demonstrated that the genetic improvement program for accurate evaluation of milch breeds of cattle and buffaloes is imperative for achieving higher economic gains as the major objective for commercial breeding is to enhance milk production and to obtain more economic return. Breeding techniques used in breed improvement programme are essentially for evolving selection criteria for dairy animals to our needs. The criterion of selection also depends on genetic parameters, superiority of the animals and the choice of trait under consideration. However, there is no suitable strategy in the field condition for assessing the productivity of each and every animal due to some burning constraints like non-availability or lack in time the availability of various interventions like breeding, feeding, management and health in terms of economic profitability, non-availability of performance recorded animals, assessing the economic profitability of the herds and diversified operations in farm enterprises. In view of these circumstances, non-availability of the productivity of the milch animal is a serious barrier for micro planning pertaining to milk production. This necessitates research to find out a way to overcome this obstacle for assessment of milk production potentiality of cattle and buffaloes. Research workers have earlier tried to predict milk production based on monthly milk yield and cumulative monthly yield for evaluation of animals, however, prediction of any day milk yield is an important area to be explored in larger perspective which may help in genetic evaluation of dairy animals under organized herds as well as field condition.

Breeding Methodology for Prediction of Milk Yield at Any Day

Research workers have tried earlier to predict the milk yield at any day using log linear Fixed Regression Model (FRM) with the help of Wood's function (an exponential function). This function can not only be used for prediction of milk yield both at farmers level as well as in a organized herd but can also be apply for prediction of days to attain peak yield, peak yield, persistency factor and prediction of total lactation yield and the model as given as :

$$\mathbf{Y_n = a * nb * exp\ (-cn)}$$

Where,

Y_n = Average daily milk yield of dairy animals in n^{th} day of each lactation.

a = constant; measures the maximum initial milk yield.

b = average slope of lactation curve.

c = rate of change of decline phase of lactation.

Breeding Methodology for Prediction of Daily Breeding Value

Under the existing dairy animal improvement program in organized herds, besides growth and reproductive performances, the elite breedable animals are selected on the basis of their breeding values, best lactation milk yield and dam's best lactation milk yield .The expected producing ability (EPA) is also used as one of the criteria for selection of female animals for test mating. For estimating the EPA, number of lactations, complete lactation milk yield, herd average and genetic parameters are also taken in to consideration. Thus, the weightage is not given for estimating the daily breeding value (DBV) of female animals for selection under the existing dairy animal improvement programme.

Today in most of the countries like USA, Canada, Netherlands, Germany, Finland , Denmark etc. the use of daily milk yield (or test day milk yield) instead of lactation yield is of more interest for predicting daily breeding values with the help of polynomial functions such as Wilmink function and Legendre function using Random Regression Model (RRM). Moreover, the advantage of using random regression model compared to lactation model is that, the former can correct more precisely the influence of non-genetic effects and simultaneously helps to improve the prediction of estimated daily breeding value which may help to identify better dairy animals and thereby improve the accuracy of genetic evaluation of animals for higher and sustainable performance. Further, estimated daily breeding values (DBV's) based on random regression model can describe more precisely the differences in the shape of lactation genetic curves between dairy animals. The daily breeding value (DBV) can be estimated in each lactation using Random Regression Model (RRM) where random regression coefficients are estimated using mathematical functions namely Wilmink's Function (WMF) and Legendre Polynomial Function (LPF), both in cubic order of fit. Wilmink and Legendre proposed a linear function of three covariates which are the well known example of mathematical model of lactation curves. Nevertheless, there is a definite trend in milk secretion through out the lactation of lactating animal which can invariably be represented by a mathematically in the form of a lactation curve consists of various phases .The phases are obtained in a consequence of initial increase in the metabolic activity of animal and thereafter fatigue in the physiological system with the advancement of stage of lactation and depressing effect of advancing pregnancy. The trend of milk production in various stages of lactation curve facilitates the estimation of persistency (the ability of the animal to continue producing milk at high level after the peak of lactation i.e. during descending phase of milk yield) leading to possibility of more accurate evaluation of dairy animals and culling of poor producers. As the lactation lengths of most of the dairy animals are being shorter therefore the lactation persistency of animals is more influenced by their lactation length.

When phenotypic production values are used for inferring the curve it is termed as 'Lactation Curve', on the other hand curves developed using genetic parameters for each day milk yield for each lactation termed as 'Genetic Curves'. The persistency within and between lactations could be judged by studying the shape of the genetic curve. The genetic persistency between days in milk 60 and 280 for dairy animal can be estimated using the cumulative breeding values on those days. Persistency of lactation is to be judged by association of genetic persistency and breeding value for persistency between lactations.

The test day milk yield records were analysed by Vohra and Chakravarty (2010) by applying the advanced Woods and Wilmink function based models in crossbred Karan Fries cattle at National Dairy Research Institute (NDRI), Karnal-Haryana. They reported different types of lactation curves among the lactating cows maintained at NDRI herd: four types of lactation curve were observed using both the functions namely standard type, reverse standard type, continuously increasing type and continuously decreasing type of lactation curves. Out of these four types, the desired type of lactation curves (standard type and continuously increasing type) showed higher lactational milk yield. According to them these desired types of curves were successfully associated with the marker (exon-2 region) of bovine leptin gene. The TT-genotype with respect to leptin gene showed the higher milk production and cows with TT genotype were more persistent compared to animals with TC and CC genotype of leptin gene in their study.

Thus, the shape of lactation genetic curves may also help further for evaluating the genetic persistency of animals and might also act as an aid to identify superior dairy animals in the breeding programme for higher and sustainable milk production.

❑❑❑

Chapter 10

Conservation of Indigenous Cattle and Buffaloes

Human interventions in the cattle & buffalo production systems during the recent decades have generally raised farm productivity. Global scenario depicts champion breeds like Holstein for milk and several others for meat, fur etc. Intensive and effective selection procedures as well as crossbreeding have variously been utilized for increasing farm output. However, competition for efficiency has reduced multiplicity and given way to uniformity, thereby leading, in general, to monoculture. Vivid example is that of Holstein for milk - this breed accounting for more than 90 percent of the milch cattle in the US. It is noted that Monoculture - in the entire gamut of agriculture and animal husbandry - does not utilize the diverse available inputs; and scarcely supports the diverse husbandry practices, or utilize the diverse climates, or survive the diverse disease spectra. Monoculture based livestock production systems are prone to diseases like bird flu, mad cow disease, and risks like trade-bias. The resulting output is less energy efficient. Under monoculture systems, the continuity of production can be easily hampered by natural disasters or even by human manipulations. On the other hand, Diversity, if present, provides innate strength to the production system and imbues the powers of long time continuity (hence, long term production). This is in contrast with short term quick gains that are generally exploitative in nature and unmindful of future generations. The gains from crossbreeding in cattle, sheep and goat in the country also need to be examined from the sustainability point of view.

Sustainability of Livestock Production

The current trends in livestock production systems show increased dependence upon new technologies, mechanization, specialization, and policies that favour industrial agriculture for maximizing production. These changes have enhanced production levels but, at the same time, have led to decline of family farms, neglect of the living and working conditions of farm labourers, and disintegration of economic and social conditions in rural areas. For livestock production, the dependence on external factors (like chemicals instead of manure, petro-mechanization instead of bull-power etc.) is continuously on the rise. A sustainable system tends to effectively produce in the current set-up but, at the same time, retaining its potential to continuously produce over time. For example, in crossbred cattle, the comparison of production from F_1 with that of F_5 can be a cause of concern. A system that can survive only by continued inflow of external input (e.g. purebred exotic bull/semen) is not a sustainable system. As a principle, a sustainable system tends to meet the needs of the present generation without compromising the ability of future generations to meet their own needs.

The largely prevalent extensive production system in the country demonstrates complete integration of crop-production and livestock-production. The two are complimentary in many respects: biologically, economically, environmentally and even socially - for the simple reason that the by-products from the crops which would otherwise add to polluting the environment, are utilized by the livestock in a fruitful manner. Similarly the by-products from livestock (dung, urine etc.) support the soil health and plant growth in an eco-friendly manner. It is such coherent complementarities that lead to sustainability of the system. The present trend of separation and specialization of crop and animal production systems has reduced their integration and has negatively affected sustainability of livestock production systems. Sustainability can be addressed by maintaining (or supporting the maintenance of) existing systems that are now endangered by one or the other limiting factor. There is a wide range of breeds and population groups available in each of the major species cattle, buffalo, sheep, goat, other livestock and poultry which have the capacity to sustain them in the local milieu in spite of the shortages and harsh conditions. Hardier breeds, in general, have lower growth and milk production potential, but are better adapted to less favourable environments.

Threats to the Self-Sustaining Indigenous Breeds

The native breeds, which traditionally formed an integral part of agriculture, are getting progressively diluted and facing degeneration. In the recent years several breeds/populations have been reduced by numbers due to neglect and lack of concerted efforts. It is estimated that, the survival of 50% of the goat breeds, 20% of cattle breeds (Figure 6), 30% of sheep breeds, and all the 18 breeds of poultry are today threatened. The focus on maximizing financial profits has alarmingly endangered the breeds which were bred for a variety of local conditions and domestic uses.

There have been several reasons leading to dilution of indigenous breeds as narrated below.

1. Introduction of exotic breeds

Since early sixties mass efforts were made to enhance the milk yield of indigenous breeds by introducing exotic germplasm (HF & Jersey) through crossbreeding with local breeds. As a result of indiscriminate crossbreeding the animal genetic resources of the country have been largely diluted and intermixes have appeared.

Figure 6: a) Punganur Cow (Courtsey : NBAGR, Karnal)

Figure 6: b) Vechur Cow and Calf (Courtsey : NBAGR, Karnal)

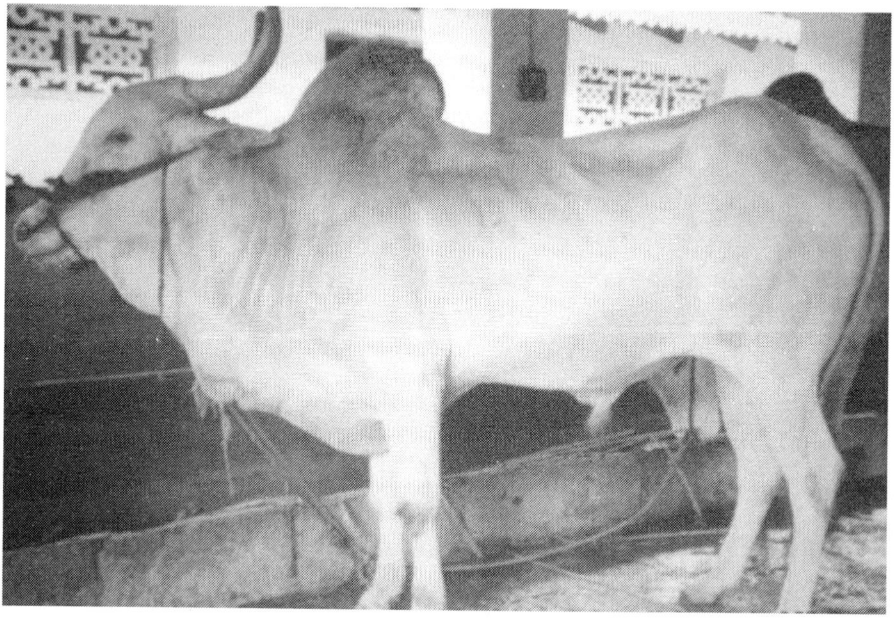

c) Krishna Valley Bull

Figure 6: Cattle Breeds at Risk (Courtsey : NBAGR, Karnal)

2. Crossbreeding by indigenous breeds

Several local indigenous breeds, which have been sustaining under their native environment, have been diluted and intermixed with breeds like Hariana and Murrah for higher productivity from their crosses.

Bhadawari buffaloes, as an example, have been known for their high fat content in milk (about 13%) and were sustainable in the ravines of Chambal River but with the popularization of Murrah for high milk yield by the State Government, their population has dwindled and this unique breed of buffalo is under threat of extinction. Likewise, there are many other native breeds, which are being replaced by other high producing indigenous breeds which are not equally efficient in the new environments.

3. Degradation of eco-systems

Indigenous breeds have been evolved to sustain under a given agro-ecosystem. The degradation of the ecosystem has caused decline in the population of adapted livestock genotypes. Mithun, being a browser in woodland habitat of broad-leaved trees/shrubs, has been thriving well in the hills of Nagaland but due to mass degradation of forests, its habitat and ultimately the population has been squeezed. Muzzafarnagri, the tallest Indian sheep breed from Uttar Pradesh, is facing extinction because the wastelands/ grasslands available in its native tract are being converted into irrigated cropping areas.

5. Mechanization in agriculture

Among the large ruminants, until recently, cattle have mainly been contributing to draught power. However, with the increased availability of mechanical power in the recent past, demand for animal draft power has decreased. Mechanization is dependent on fossil fuels which are non-renewable and non-sustainable, besides being a drain on the national exchequer.

Need for Conservation of Sustainable Indigenous Breeds

The native breeds are better adapted to the local agro-ecological conditions of the region and are capable of producing on almost zero input system. Even the long-term sustainability of breeding efforts may depend on the continued availability of the genetic variation. The indigenous breeds need to be conserved for a number of reasons as follows:

1. Crossbreds are more productive as compared to native breeds but their tendency to wilt under Indian conditions of low input and harsh climate, susceptibility to tropical diseases, drop in production levels beyond F_1 warrant conservation of indigenous breeds for future use. Production from indigenous breeds is more sustainable.

2. No organized efforts have been made to improve the genetic potential of indigenous breeds which are capable of thriving under the existing low input and poor management conditions, and hence, are more sustainable.
3. Usefulness of various types of indigenous animals has not been fully explored, nor their sustainability fully examined.
4. The non-renewable sources of energy are bound to exhaust sooner or later. If this does happen, then we may have to fall back on our animal wealth for providing draught power and hence we cannot take risk of letting these breeds go extinct. Bulls and cows represent a self-generating and self-sustaining system without requiring external support.

Conservation Strategies

Broadly, there are two means of conservation i.e. *in situ* and *ex situ*. Conserving in the form of live animals that exist in nature is *in situ* conservation. In comparison, *ex situ* conservation is achieved by storage of sperms, oocytes, embryos, DNA, Somatic cells and embryonic stem cells. The most appropriate means of conservation is in the form of live animals. However, if the population is too small (critical), *ex situ* conservation must be utilized to save the breed from extinction. In order to implement conservation of animals, the following activities have been initiated and need to be taken up in a big way.

1. *In-Situ* conservation strategy

Explicit efforts to select males from superior dams in farmer's conditions and making wider use of the selected best bulls and also preserving their semen are necessary. The process has been initiated for some of the breeds by NBAGR. A few Gaushalas are also carrying out the task of conserving and improving local cattle breeds.

2. Data bank strategy

Maintenance of a database containing all relevant breeds, population census, and ecological data is essential for designing and implementing conservation strategies. Several agencies are engaged in generation and dissemination of data/information on Animal Genetic Resources. A useful body of knowledge has already been generated/gathered at NBAGR and at other locations. Mechanisms for effective sourcing and pooling of information through networking and data-warehousing need to be developed. International Council of Animal Recording (ICAR) is involved in data recording and International Sire Evaluation Programme.

3. Gene bank strategy

Semen from indigenous breeds should be cryopreserved for use in future. Ideally, sufficient doses should be stored at least at two locations remote from each other. The preserved material should be periodically evaluated and put to use. Semen from proven / good bull should be cryopreserved.

4. DNA bank strategy

Genetic material can be preserved in the form of DNA fragments under cryogenic conditions. This has the advantage over storage of live cells as it is economical, occupies less space and there is no chance of spread of diseases. Within and across different countries the storage of DNA has been made feasible. The genome maps of all farm species are not yet available (except cattle); hence it is not a normal method of preservation at present. Nevertheless, DNA fragments with the "effective gene part" can be preserved for possible future use in transgenics.

5. Somatic cell strategy

With the advent of Dolly sheep, somatic cell technology has received a great fillip. In future it may be possible to produce a live animal from stored somatic cells. This possibility is very important since the protocols for collecting somatic cell samples are less demanding and inexpensive than for collection of spermatozoa and embryos. Recently in 2010, "Garima" buffalo has been cloned by NDRI Karnal.

Summary

There is a wide range of breeds and population groups available in each species of livestock which have the innate capacity to sustain themselves in spite of harsh local conditions like high altitude, high salinity, desert or drought and still be able to survive as well as produce. Often an imported breed is unable to produce or even maintain itself under such conditions. Such sustainable systems need to be maintained for continued production and as future genetic insurance systems.

❑❑❑

Chapter 11

Genotype & Environment Interaction

It is well known that the performance of animal depends not only on the superior genotype but also the environment in which they are born and reared. It is a common opinion among animal breeders that the relative merit of various breeds or strains also depends on the external conditions (environment) to which they are exposed in addition to their genetic make up. In other words there may be an interaction between genotype and environment (GXE interaction). A familiar example is the performance of Zebu and European cattle in temperate and tropical climates respectively. If GE interaction is significant, it will bias the sustainable performance in animals to the extent that it reduces accuracy in predicting the production level.

Introduction to GXE Interaction

The phenotypic expression of any quantitative trait is the joint effect of genetic and environmental effects on its development. The best possible genotype will not exhibit its full potentiality unless the proper environment is provided. Likewise the best possible environment alone is not sufficient to develop a superior organism unless the proper genotype is present in the organism. Thus, one environment may permit the expression of a genotype while others do not. This differential response of different genotypes in different environments is called genotype environment interaction. In terms of gene action one may assume that some genes may have effects that differ from environment to environment. Such interactions may cause decreased performance when a population selected under one environment enters in a new environment i.e. a regression in performance, that Dickerson (1962) termed as "Genetic slippage" which is the reduction of breeding value when population perform in environment other than that in which it was selected. A naturalist considers the GE interaction as differential adaptability of different genotypes (species, breeds or races) under different ecological conditions. To a statistician and animal geneticist, the GE interaction means that the genotype and environmental variations do not combine their effects linearly or additively i.e. a phenotype is not simply the sum of the genotype and the environment but another component of variation (GE interaction) is associated and thus the mathematical model for phenotype (P) can be taken as:

$$P = G + E + I_{GE}$$

Where,

I_{GE} indicates the genotype environment interaction

At this point of discussion a clear understanding about the word "environment" demands special consideration. The environment of an individual is made up of all the things other than the genotype of the individual that effect its development. That is to say, environment is the sum total of all non - genetic factors external to the organism. Comstock & Moll (1963) distinguish two kinds of environment micro and macro. Micro-environment is the environment of a single organism as opposed to that of another growing in the same time and almost in the same place. Micro-environmental differences occur even when individuals are apparently treated alike. On then other hand, environments that are potential or realized within a given area and period of time is referred collectively as macro-environment. A macro-environment is thus a collection of several macro-environments. Different locations, climates, nutrition, disease prevalence and even different managemental practices are examples of macro environmental differences in domestic animals. It is to be noted that the effect of micro-environment on an organism as well as its interaction with genotype is usually very small. Macro-environmental deviation and its interaction with genotype can be isolated and tested for significance.

Statistical Interpretation of GXE Interaction

An interaction between genotype and environment gives rise to an additional component of variance which can be isolated and measured only under rather artificial circumstances. We may replicate genotypes by the use of inbred lines or F_1 and replicate specific environments by controlling factors such as nutrition and temperature etc. Then analysis of variance in two-way classification of genotype environmental variance, and the variance is attributable to genotype environment interaction. The existence of GXE means that the best genotype in one environment may not be the best in another environment e.g. the breed of cattle with high milk yield in a temperate climate is unlikely to have the highest yielder in tropical climate. This has an important bearing in breeding policy, particularly with respect to sire evaluation. This give rise to a question in mind that if a sire is ranked as best under best condition of feeding and management in one farm, will the achievement be carried over in later generations or when the same bull is used is poorer conditions under which the majority of animals are required to live in India, whether the results will be same? By regarding performance in different environments as different characters with genetic correlation between them, we can solve the question raised above from the knowledge of heritability (h^2) of different characters and the genetic correlation between them (Falconer, 1952). If genetic correlation is high, performance in two different environments represents nearly the same set of genes. If it is low than the characters, to a greater extent, are different and high performance requires a different set of genes and it will be advantageous to carry out the selection in the environment in which the population is destined to live unless the h^2 or intensity of selection in the other environment is very high. This is the theoretical basis but the results of the experiments that have been carried out do not appear to be fully in agreement with theoretical expectation and this suggested that other factors not yet understood are probably operating (Haldane, 1946; Falconer and Lady Szewi Ski, 1952). The total genotype and environmental interaction appears to be due to non-additive genetic effect.

Classification of GXE Interaction

To be more illustrative, two basic types of GXE relationship have been recognized viz. Linear and nonlinear.

1. Linear/Additive relationship

The linear relationship is indicated when the phenotypic differences between genotypes remains unchanged / constant in different environment.

2. Non-linear / Non-additive relationship:

This is actually known as GXE interaction and is of two types:

a) When quantitative differences in the phenotypic values changes with change in environment but the rank order do not change. This involves change in the size of differences among genetic groups but do not change the rank of groups.

b) When the distinguishable characteristics are a reversal of phenotypic rank orders as the genotypes are moved from one environment to another. These involve changes in rank order among genetic groups. The difference among individuals (genotypes) as it exists in a particular environment may be classified into two intra and inter population differences. The intra-population genotypes include small differences genotypes forming a continuum for genetic merit within a population like single herd of dairy cattle or flock of sheep. The inter-population genotypes include large genetic differences like several lines, strains, breeds or crosses. In view of the above distinction among individuals and differences of environments viz. Micro and Macro, McBridge (1958) gave the following four types of GE interaction:

Environment/ Genotype	Micro Environment (Small Differences)	Macro Environment (Large Differences)
Intra Population genotypes (Small Differences)	Type A	Type B
Inter Population genotypes (Large Differences)	Type C	Type D

Out of these four types of interactions, the type B interaction i.e. intra population Macro-environment is of special significance in concerned to sire evaluation programme, since their existence necessitates the practice of selection of the sire within the environment for which the animals are required. This type of interaction becomes more important when selection is with in a breed or strain carried out under different environmental conditions. Sire-ration, sire-herd, sire-year-season, sire-region are examples of this type of GXE interaction.

Type D interaction includes when breed differences are tested under macro environmental changes. These are important in dairy cattle breeding. If this type of interaction is important, it will then decide the relative merit of lines, breeds in different environments and will need to evaluate the adaptability of different breeds under different environments. The approach to the problem of GXE interaction in sire selection and evaluation programs in dairy cattle has been largely of estimating interactions of sires with herd production levels, locations, nutritional levels and year-season. Sustainable performance in dairy animals may depend on various combination of Genotype X Environment.

GXE Interactions

Sire X Production-Level of herd

Better herd production reflects better management. This type of interaction may alter the ranking of sire. Consequently, the bulls selected on ill managed herds, may not prove themselves worthy, if used on high production level. However, many workers have concluded that the sire ranking will remain more or less same at different levels of herd production. Mc-Danial and Corley, 1976; Normon *et al.* 1972; Rao, 1979; Rao and Taneja 1982, concluded that the sire x herd level interaction for milk yield in Holstein X Sahiwal crossbreeds to be non-significant. Van Vleck (1963), however, found that the genetic progress in higher herd level was comparatively lower when the bulls were evaluated in low levels of production and suggested that it will be advantageous to evaluate the bulls at better herd levels, so that they will prove worthy at most of the production levels.

Sire X Location

The differences among regions / farms arises mostly due to the climatic variation in physical environmental parameter (temperature, humidity, wind velocity, altitude etc.) in addition to management variations. The studies of Peterson (1975), Taneja and Garg (1976), Rao (1979) and Chauhan (1985) revealed non-significant sire X region/ farm interaction. On the contrary Sahota and Gill (1983) computed sire indices (weighted herd mates comparison) for 300 day milk yield using data of Friesian X Sahiwal crosses of eight military dairy farms. Although, the rank correlation were non-significant, the results suggested the need to carry out comprehensive studies on GXE interaction both in farm as well as field conditions to ascertain their role.

Cunninghan and O' Byrne (1997) (cited by Syrstad, 1990) compared progeny tests of bulls used in Ireland with tests of their sires in Britain and estimated a genetic correlation of only 0.6 between milk yields as expressed in the two environments i.e. a large GXE interaction. Moulick (1978) and Buvanandan and Peterson (1980) also reported significant sire x region interaction in milk yield of dairy cattle.

Sire X Nutrition

The sire X ration interaction was found to be non-significant for milk production (Lamb, 1964). The studies on identical twins also revealed negligible genotype X ration interaction for milk yield and growth traits (Syrstad, 1978). In contrast Richardson *et al.* (1971), Lamb *et al* (1973), Holroyd *et al*(1979) and Clrenek and Flak (1981) reported significant genotype X nutrition interaction in production traits of dairy cattle. Taneja and Bhat (1978) while trying to separate out the additive effects from heterotic variances in Holstein X Sahiwal grades observed that the partial regression coefficient for body weights were significant indicating the involvement of non-additive gene actions. Heterozygosity was also important for age at first calving, and first calving interval. Milk production has shown the effect of heterosis positive but of non-significant in nature. Thus, this type of interaction may be important where sires are selected on performance of daughters on higher levels of feeding, are to be used in herds fed mostly on roughages. In presence of major differences between the feeding regimes, in terms of quality and quantity, of herds on which sires are tested and to be propagated, sire x nutrition interaction should not be ignored.

Sire X Year-Season

Rios and Branton (1973) reported that sire and year season interaction for milk and fat yields and conception rates were not major sources of error in sire evaluation. Sire by season of calving interactions for complete and part yields were non-significant accounting for less than 1% of the total variance. Kelleher *et al* (1967) observed that bull by herd-year-season interaction for milk production would not cause any appreciable bias in sire selection programme.

Summary

For proper evaluation of sustainable performance in dairy animals, a detailed knowledge of the performance of different genotypes in different environments is needed. From the past experience it is evident that GXE interaction will not hinder production unless these are significant. Reviewing the results above, it appears that although the GXE interaction is not a major source of variation in sustainable performance, selection of sires in most of the cases and ranking of sires however cannot be ignored totally, especially when the environmental levels are wide apart. For sustainable performance it is important that the high expenditure in evaluating a sire has to be justified by its extensive use. Over several herds across the regions, it will, therefore, be desirable to study the significance of GXE interaction before appraising the breeding value of the sire, especially in field progeny testing programmes where the environmental variations are large enough. Allotting sires to many herds of different sizes and ensuring large numbers of progeny per sire could minimize the magnitude of GXE influence. This will evaluate the sires most effectively over the herds distributed in many locations. Although the daughters of the sires kept under better management conditions shows better performance but the ranking of the sires did not change under different management situations. However, a caution is needed to conclude that such interactions do not exist as only small part of vast array of genotypes and environments have been studied. Therefore, these interactions could be quite important where environmental and management situations differ significantly and thus there is a need to develop the suitable models for the analysis of performance data over the different herds under various breed improvement programmes for proper evaluation not only for males but females as well.

□□□

Chapter 12

Sustainable Breeding with Assisted Reproductive Technologies

Artificial Insemination (AI)

AI remains as one of the most important assisted reproductive biotechnology. The history of research into Artificial Insemination (AI) is over two centuries old and its commercial application now spans 90 years. It is appropriate to reflect on the contribution of this powerful method of gene disposal. The three cornerstones for its application are. It is simple, economical and successful. The remarkable progress made in other assisted reproductive technologies does have the potential to rapidly generate offspring but the challenges for any of these reproductive technologies to attain widespread use to match AI in being simple, economical and successful is inconvincible. Genetic advancement in the livestock population and particularly in the dairy cattle & buffaloes has relied on two processes, use of the top genetic merit bulls and the selective rearing of calves of high breeding merits as replacements. To this end, AI has remained the main vehicle for the rapid disposal of valuable genes and it has been the method of choice of dairy farmers around the world to improve the genetic quality of their stock. This steady level of genetic progress in dairy cattle is primarily due to advancement in semen technology and rapid access of AI to established favourable genes in the cattle population. The mains drives for uptakes of AI are, cost of the semen, cost of the insemination, and the success rate of the process. The genetic contribution of a sire is described by Foote (1998) as

Contribution of a sire to genetic improvement = Number of progeny per sire X Genetic superiority of the sire

The number of progeny per sire will be determined by

1. Total sperm output of the bull.
2. The number of sperm used per insemination.
3. The percentage of cows calving to a single insemination.

This can be represented as

Number of progeny per sire = Number of sperm per sire/Number of sperm inseminated per cows X proportion of cows calving to a single insemination.

These principles effectively determine the number of bulls required for servicing a dairy cow population. On average, at the current technology level, the number of doses possible for a given sire ranged between 1,00,000 and 1,50,000 if used exclusively as frozen semen. However, seasonal breeding pattern in tropical countries and vaccination limit the production of frozen semen in actual terms.

A value analysis of AI requires it to be compared with other mechanism of gene dispersal such as natural mating and embryo transfer. The main driving force in any such value analysis is that the objectives need to be clearly understood. This varies between different environments. Various aspects of artificial insemination technology have been fairly standardized. It has now become a practical technology in commercial dairy cattle programs in both developed and developing countries, covering a worldwide total of about 50 million first inseminations (Chupin and Schuh 1993, Chupin and Thibier 1995).

In India, where well structured dairy development programme have been established with cattle and buffalo, artificial insemination (AI) procedure has played a valuable role in facilitating appropriate genetic change in animal populations, being linked to intensive male selection. In developed countries, AI system is increasingly being developed and used in the breed development of livestock genetic resources. AI technology offers certain specific advantages, especially widespread use of outstanding males and dissemination of superior genetic material; progeny testing under environment and managerial conditions to improve the rate and efficiency of genetic selection; accelerated introduction of new genetic material by import of semen rather than live animals and thus, reducing the international transport costs; enabling the use of frozen semen even after the donor is dead; and reducing the risk of spreading sexually transmitted diseases. Long-term semen storage, without loss of viability, for use in AI, is another valuable technology for promoting conservation of endangered breeds of farm animal species, although this technology has the disadvantage of preserving only half genotypes and requiring secure cryo-preservation facilities. However, AI will remain the most cost-effective method to disseminate these genes. It is more than likely that the combination of AI with new and emerging biotechnologies will form a strong base on which various selections is based.

Semen Storage

For the purpose of conservation, proliferation and development of germplasm, semen banks have been established throughout the world. Functionally semen bank/ bull station involves the activities of production of semen (Figure 7 & 8) from well-managed and genetically superior bulls, semen evaluation processing, storage and supply for its subsequent utilization in the field (Figure 9). Semen bank is the nucleus of the artificial insemination organization. Sometimes it is part of a farm (dairy farm /bull mother farm) as well. Main objective of Semen bank is quality frozen semen production so as to ensure better fertility and conception rate, and faster genetic progress. Progress in the science of low temperature biology has made it possible to store a wide variety of living cells for a prolonged period at a very low temperature. Developments in successful freezing and storage of semen further stimulated the adoption of AI especially for cattle, leading spermatozoa to a state of quiescence by reversible reduction of metabolic activity at the time of collection and subsequent method used for inactivation, revival and utilization of semen. The processing of semen starts with the bull and ends with the cow. This necessitates a high level of cooperation and coordination among the person engaged in management of bulls, handling semen both in laboratory and field, and in collection, assessment and interpretation of data (Madan, 2003). Storage of semen in the frozen state implied a complete arrest in the developmental and metabolic process within the sperm cell. The cell need to be held in a state of suspended animation until required for fertilization. Storage of frozen semen is expensive as liquid Nitrogen and well insulated containers are essential. Freezing of semen has now become an established commercial practice especially in all domestic animals.

Frozen semen offers a wide range of benefits as given below:

1. There is virtually no wastage in utilization of frozen semen.
2. After required numbers of doses have been collected from a bull the animal can be disposed off, thus making the programme economical.
3. Other benefits are the use of bulls of better breed with high production potential, use of semen over a longer period, services available at door step of animal keeper, control of spread of reproductive diseases, implementation of progeny testing and other livestock improvement programmes, preservation of semen from a number of bulls simultaneously in one container and transport of frozen semen being possible even across the continents.

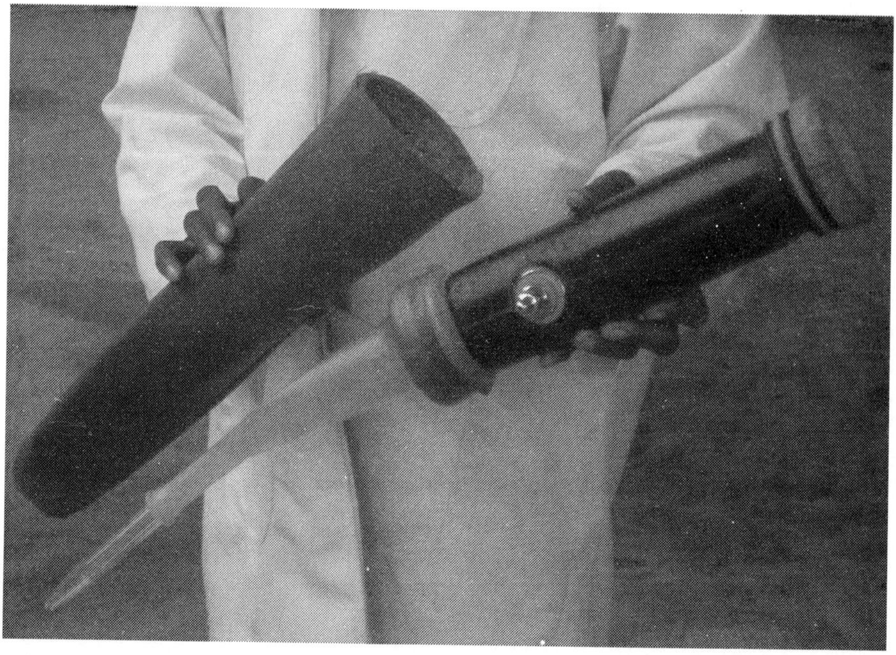

Figure 7: Artificial Vagina Assembly used for Cattle & Buffalo Bull

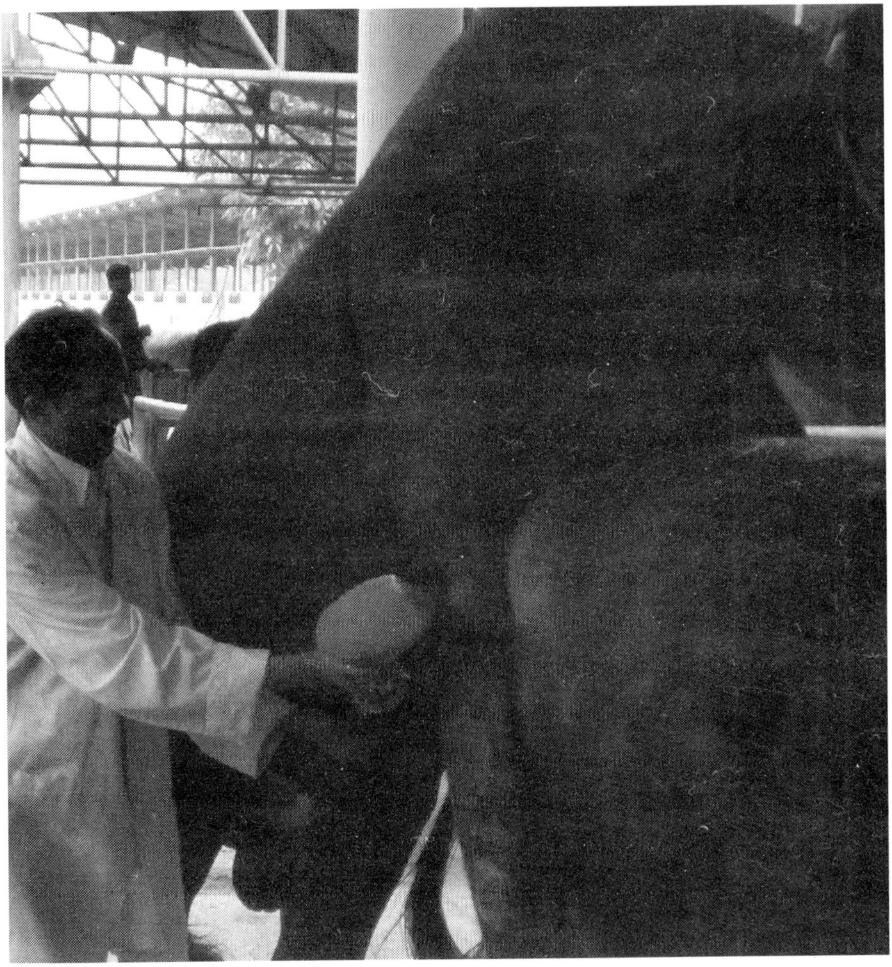

Figure 8: Semen Donation and Collection

Impaired Fertility of Cryopreserved Semen

Cryopreserved mammalian semen is generally acknowledged to have an impaired fertility in comparison to that of fresh semen (Watson, 2000). The reduction arises both from a lower viability post thaw and sub-lethal dysfunction in a proportion of the surviving population. The reasons for this decrease in fertility are cold shock susceptibility, osmotic stress and intracellular ice crystal formation during freezing and again during thawing (Mazur, 1984). Also cooled boar semen experiences "capacitating like changes", which may be partially responsible for the reduced fertility (Green and Watson, 2001).

Figure 9: Semen Straw Filling Machines

The discovery of glycerol as cryoprotective agents introduced a system of freezing semen that until recently has seen remarkably few changes. What is even more axiomatic of semen freezing is that even with the most refined freezing procedures, controlled freezing conditions and other diluents modifications; the best cell recovery after thawing is just over 50%. Over the last 50 years, a huge body of knowledge has been accumulated on the physical stress to sperm membrane during the freezing process. The volumetric changes, oxidation of sperm membrane lipids, damage to selective permeability mechanism of the membrane are all well understood, yet imaginative methods to combat these changes have not been devised. In recent years, the oxidative damage that the spermatozoa may suffer is receiving increasing interest. It has been shown that semen samples with highest viability after freezing and thawing, were also characterized by high superoxide dismutage activity (SOD) and high proportion of selenium dependant glutathione peroxidase (GSP-Pe-Se) activity before freezing. Within anaerobic or even a partially aerobic system, the production of reactive oxygen species (ROS) is unavoidable; the damaging ones being the superoxide anions (O_2^-), peroxide (H_2O_2) and the hydroxyl free radicals (OH). Peroxide is the most pernicious of all the pro-oxidants formed. It has been hypothesised that extenders containing egg yolk enhances the effect of H_2O_2, due to the presence of iron and other transition metal according to Fenton–type reactions.

Oxidative damage

Mitochondrial DNA and membrane architecture may be a factor of major importance to explain the impaired fertility and motility of cryopreserved semen and since the mitochondria of sperm mid piece generate energy to support motility, change in the mitochondrial membrane potential could be a good indicator of functional impairment of cryopreserved semen.

Antioxidant incorporation

It has been demonstrated that alpha tocopherol (Vit.-E) prevents oxidative damage on boar sperm membrane in semen maintained in liquid state (Cerolini et al., 2000), in buffalo semen (Raina et al., 2003). The ability of Vit.-E to inhibit lipid peroxidation in boar semen induced by means of sodium ascorbate and $FeSO_4$ has been reported.

Table 12: Format for Record Register to be maintained at Bull Mother Farm

S.No	Species / Breed	Bull No.	Ejaculate No.	Quantity (ml)	Colour	Mass Motility	Klet Reading	Diluents Used	Diluents Ratio	Sperm Conc.	Motility After Dilution	User Name & Remarks
1												
2												
3												
4												
5												
6												
7												
8												
9												
10												

Challenges in Artificial Insemination

The semen delivered to the farmers should have an intrinsic property of being breed / programme specific (see Table 12) of high genetic merit, with productivity or superiority of a character better than the breed/animal the farmer have. Semen should have high fertility and should be free from diseases and infections. Every male calf born in the country, normally qualifies as a bull (which should not be) as many of them are used for actual service. Even in Semen banks we have a limited number of Progeny Tested Bulls, when in fact only bulls, which have proven superiority, should have been used in A.I. programme. The 219 Semen Production Centres and 139 Frozen Semen bank in the country are very marginally able to contribute to the total needs of semen that too often under doubtful quality of breed potential and fertility. The semen production capacity of a bull (buffalo 10,000 doses and cattle 24,000 dose annually) demands a high number of bulls framed into service for the total semen needs. The adult female population of 40 million buffaloes, 57 million cattle and about 10 million crossbred female according to an estimate will requires about 15,300 buffalo, 9500 cattle and 2000 crossbred bulls in numbers which are too large.

The challenge therefore is to identify our demands in a phased manner in consonance with the required demands, breed application and the breeding policy. The breed viability among both cattle and buffaloes and plethora of nondescript animals adds another dimension to the challenge. Since, a clear cut policy has to be evolved for utilization of semen on these animals on different ecological regions and on breeding tract basis. The production excellence of a livestock breeds can only be exploited if the best of the animals are used as bull mothers and calves are produced through nominated mating. Such elite animals are least to be found in Farm or in Government institutions. Bull production should therefore be essentially connected to the field evaluation of excellence involving identification and recording and linked to use of best semen on field animals and an incentive must be given to the farmers. This is the only sure and economic method for having bulls for different breeds produced from diverse genetic group. A challenge is to have a thorough assessment of the numbers of different indigenous, buffalo and cattle breeds, evaluating the potential of a bull already in use and producing bull calves for future use.

Bulls' Role in Genetic Improvement

The result of selecting superior animals as the parent stock will change the mean performance in the progeny generation. The identification of superior individuals is a complex process starting with the birth of the animal and is completed in different stages, conveniently grouped as initial selection (or) phenotypic selection which is based on physical attributes and the breed characteristics. Final selection (or) genetic selection of bulls is based on its breeding value (the sum of the average effects of the genes possessed by the individual for a particular trait) breeding value is estimated based on the performance of bulls relatives. In dairy cattle when AI is employed as means of breeding, genetic information is passed on from generation to the next through the paths and in the proportion (in percentage) shown below (Robertson and Rendel, 1950).

- Bulls to bred bulls = 43%
- Bulls to bred cows = 18%
- Cows to bred bulls = 33%
- Cows to bred cows = 06%

The selection of the dams and sires of future bulls thus accounts for more than three fourth of the total expected genetic change. Very little is done in our country for systematic selection of bulls used for AI. Even at present for selecting a bull from any one of the organized farms, all we do is to check the dams' yield.

Quality Control in Semen Production

The fertility potential of cryopreserved semen can be influenced by numerous factors: the most important of which is the quality of semen obtained in the raw ejaculate. In addition to inherent variation in semen quality amongst bulls, management, nutrition and environmental stressors can have a transient influence on spermatogenesis for weeks to months. Because of the numerous interactions between individual components of the cryopreservation process each AI organization must develop and continually revalidate semen collection and processing protocols using acceptable viability criteria and final thawing and handling procedures that are consistent with recommendations.

Once the frozen semen leaves the control AI centre there are numerous means by which semen can affect viability and fertility in a negative fashion. Comparative research data indicates that relative magnitude of viability damage incurred to cryopreserved semen is greatest when semen is warmed and the re-cooled to liquid nitrogen temperatures. This is followed by improper thawing procedures and lack of post thaw thermal protection. It is essential for each inseminator to ensure that the thawing protocol of the organization, which processed the semen, is being followed exactly as described with no deviation or modifications or short cuts. The dimensions of the AI have been broadened tremendously by technological developments, first, permitting long term preservation and storage of frozen semen through freezing and then permitting development of systems for world wide distribution of frozen semen.

The magnitude of the infection potential of one bull whose semen is used in a frozen semen program is very large. The regular testing of semen donors under official veterinary supervision has been adopted by government's world wide as a means of avoiding the spread of pathogens and reducing excessive contamination of semen by ubiquitous bacteria. National standards for semen production and distribution are usually based on regulatory programmes to ensure that important diseases are identified and appropriate tests are applied to all sires entering and residing in AI centres. These programmes take into account the national health status as well as the health status of the herds and flocks of the semen donors. In interpreting health status prime consideration include the sensitivity and specificity of tests particularly when they are applied to individual animals and risk of latent infections. Testing programmes need to be continually improved and updated as new information on the epidemiology; pathogenesis and control of traditional diseases become available. Genetic disorder occurring in animals can be mainly attributed due to intense selection for improved animal performance, since this may inadvertently increase the frequency of undesirable traits of improved sires. The genetic diseases in cattle being inherited as autosomal recessive traits, the carriers generally give no outward indications, and undesirable trait can be spread widely and covertly. The application of molecular biology has revolutionized the understanding of genetic basis of disease susceptibility of farm animals. The major breakthrough is the establishment of PCR technology. In many countries PCR assay was implemented to screen bulls entering artificial insemination studs. Karyotype studies are also found equally important for detecting certain defects. Except few, many chromosomal translocations have been not linked with a disease but it is becoming an accepted practice not to use such animals for AI.

Implementation of standards for quality Frozen Semen production in the country is crucial to livestock improvement. Though several established centres have an acceptable record of the results but there has been no attempt for harmonization of protocols so that we have uniformity in terms of strength of one station being shared by others. Quality control in frozen semen centres leaves much to be desired. A total quality assessment should be in terms of the guidelines given by O.I.E for disease screening and continuous evaluation of the semen. The Microbial test of the semen should be routine and frozen semen should enter the semen bank only after microbial screening and after quarantine/test analysis.

Major Constraints in Artificial Insemination

Success of AI is connected to a chain of events, the strength of which lies in its weakest link. Major constraints identified include stray bulls, absence of cold chain, semen of poor genetic merit, poor semen quality, poor fertility of bulls, disease prevalence among the bulls, lack of training and motivation of the inseminators and limited AI infrastructure. However, field staff and the scientific/Lab personnel differently perceive these constraints. While training and motivation of the inseminator is correctly perceived as the first priority by the field staff, technical personnel consider semen of poor genetic merit as the major constraint. An effective Frozen Semen production and delivery system can eliminate several of the constraints.

Embryo Transfer Technology

Recent reproductive technologies like ET and IVF are the only means by which elite females could be multiplied faster. Considerable progress has been made in methods of recovery, storing and implanting cattle and buffalo embryos. In India also, techniques and protocols of ET technology are being standardized in cattle, buffalo, sheep, goat, camel, and other species of animals. Embryo transfer and other associated reproductive technologies have been successively used for rapidly multiplying the populations of elite breeds of cattle (Willet et al., 1951; Thibier 1996), buffalo (Drost et al., 1983; Madan et al., 1993; Misra 1993; Misra et al., 1999), sheep (Holm et al., 1996; Thibier 1996), goat (Armstrong et al., 1983; Pawshe et al., 1994; Thibier 1996), horse (Riera and McDonough 1993, Squires et al., 1999) and pig (Galvin et al., 1994; Hazeleger and Kemp 1999).

Advantages of MOET

(i) Rapid expansion of rare and/or improved exotic/local genetic stocks of farm animals and even elite female animals

(ii) Reducing the cost of international transport by shipping embryos rather than living animals in which case quarantine restrictions would also apply.

(iii) The rapid replacement of existing genotypes by using ET/MOET rather than grading up through repeated crossing.

(iv) The possibility of increasing the twinning rate by combining AI with the transplanted embryo (Cunningham, 1999) and also

1. Pre-natal sex determination
2. Control of sexually transmitted diseases
3. Using biological diversity of embryos
4. Environmental adaptability as compared to transport and establishment of imported animals.

In combination with other animal biotechnological procedures, ET/MOET techniques can accelerate herd development and animal conservation especially of rare genetic stocks. Embryo splitting technique and transfer of split embryos have been standardized by National Institute of Immunology (NII), and the first bovine split embryo derived calf was born in November, 1988 (Raghupathy and Hasnain 1991; Ramachandran, 1991). Such technologies enabled the production of 150 cows from an elite cow, as opposed to ten cows that she would have normally produced in her lifetime in conventional husbandry. Scientists at NII also succeeded in producing a 100% pure Holstein-Friesian calf, using a local stray cow as surrogate mother (Raghupathy and Hasnain 1991, Ramachandran, 1991).

The main use of ET is the production of bulls for progeny test through AI. However, now that the costs of embryo transfer are decreasing, production of replacement females within a herd could become a possible use of ET as well. The true economic value of ET should determine the extent to which ET techniques are applied. Until recently, high costs associated with embryo transfer have restricted its use to farms where a good share of income comes from sale of breeding stock. To be useful to commercial dairyman, extra income from production must pay all costs. For production of replacement females, many factors influence profitability. From the benefit side, those factors include: fraction of selected donors, merit of mating sire, discount rate, planning time horizon, etc. From the cost side, economic factors include equipment costs, recipient costs, labour costs, semen costs, etc. Cost benefits comparison for three different scenarios using ET and/or AI.

The scenarios are:

1) Whole herd AI,
2) Whole herd ET,
3) Partial ET and AI.

- The results of a study indicate that it is indeed possible to use embryo transfer as a tool to increase income from a commercial dairy herd. However, to experience profit from ET, it is imperative that only the females with the top EBVs for total economic merit (e.g. LPI or preferably TEV) be used as donors, and the lowest animals be used as recipients.
- Partial-herd ET (i.e. combined use of ET and AI) was always more profitable than whole-herd ET, provided that AI is used for cows with intermediate EBVs total economic merit (e.g. LPI or TEV).
- If the maximum number of flushes per donor is set to five, highest profits are achieved when flushing the top 2 to 3% cows for total economic merit.
- Profitability of an ET-program heavily depends on the success of ET and the costs involved with ET.

Success of ET

Number of viable embryos per flush and embryo transfer conception rates is very important. As these results improve, the costs of producing an embryo decrease dramatically.

Cost of ET

ET has been priced more affordably over the last few years. When farmers perform ET themselves costs are even lower, but even slightly poorer success rates can easily eliminate gains from lower ET costs. Therefore, the aim for farmers should be to get higher ET success rates with the lowest possible costs. Lower costs per embryo improve the possibility of flushing more cows and achieving higher profits.

Embryo Storage

Freezing of semen and embryo is a standard practice especially in cattle. Preservation of embryonic stem cells could represent an important method of genome conservation. With *in vitro* Maturation and *in vitro* fertilization (IVM/IVF) Embryo Production, it is now possible to adopt a fully *in vitro* system, whereby the immature oocytes are matured *in vitro*, followed by fertilization with *in vitro* capacitated spermatozoa and the newly formed zygotes are cultured in suitable media for development upto the transferable stage. The births of such *in vitro* animals have been reported in cattle, goat, pig and buffalo. However, recent development of low invasive ultra sound guided transvaginal oocyte retrieval (TVOR)/ oocyte pick up (OPU) has permitted production of more embryos than might be possible by standard ET practice. TVOR also allows repeated collection of oocytes from endangered species of livestock or livestock of high economic importance in order to propagate such genetic resources in much faster way. Freezing of high quality oocytes and semen would provide germplasm for undertaking important future livestock improvement and germplasm conservation programme.

Sexing Semen and Embryo

It is possible now the sorting of semen, one sperm at a time, into males and females, using staining procedure and detecting by laser beam with the help of standard Flow Cytometry equipment. Thus, sperms with an X or Y chromosome could be used to produce male or female embryos /animals. Embryos can also be sexed. Embryo sexing is done at early embryonic stage with use of sex specific DNA probes.

Cloning

The successful cloning in domesticated animals has been achieved, using early embryonic source material, nuclear transplantation, embryonic cultured cell line and somatic cells of adult animals (Dolly sheep). Cloning using somatic cells offers opportunities to select and multiply animals of specific merits. Cloning holds the promise of bypassing conventional breeding procedures to allow creation of thousands of precise duplicates of genetically engineered animals or other animals in a single generation. In remote areas, where sampling and storage of adequate samples of semen and embryos is not practical, one could use cloned samples from diverse animals for conservation of the available genetic diversity of such threatened genetic resources.

Cloning may eventually prove to be much simpler and effective means of conservation of breeds. The rapid multiplication of superior germplasm can be attained through these techniques of embryo transfer. Several National Institutes viz., Indian Veterinary Research Institute, National Dairy Research Institute, National Institute of Immunology and National Dairy Development Board, have already initiated research and development programmes using these biotechnologies. Assessment of impact of these technologies on our cattle and buffalo populations has yet to be made.

Summary

The genetic improvement programmes can attain greater heights with the help of assisted reproductive technologies. The sexed semen and embryos are the new addition and emerging technology in India. Some states of India like Haryana, West Bengal are on their way to generate sexed semen. Sexed semen and embryos can lead to higher genetic response rate in genetic improvement programmes. Artificial insemination is and will remain the most extensively used and an important tool in sustainable breeding of cattle and buffaloes.

Part-2

Feeding & Management Interventions in Sustainable Breeding

Chapter 13

Feeding Requirements in Cattle & Buffalo

Introduction

Unlike western world, the bulk of milk production in India comes from small and marginal farmers, who raise small number of animals primarily for draught purpose, mostly on crop residues as the by-products from grain farming. An assessment of the major constraints in animal production across the world's developing regions, identified nutrition as the most important factor (ILRI, 1995). The problem of inadequate feed supply is all the more acute in India, in view of the large bovine population of the country. There is also an increasing pressure on land, from ever increasing human population. Due to this arable land is shrinking and natural resources are being depleted, none of which is conducive to dairy development. Although with the self-sufficiency in the food grains, the quantity of feed resources in India have increased by almost 70 per cent from 1954 to 1996, yet the availability of feed / unit has remained almost the same, due to corresponding increase in animal population. The gaps between nutrient requirement and availability have also been estimated as 46.2 percent with DCP and 44.2 percent with regard to TDN for bovine population of the country (Jam *et al*, 1996). The area under fodder production has decreased. Despite of the fact that the crop residues availability has increased and compound livestock feed production has also increased still, as mentioned above, we are not able to meet the nutrient requirement of our livestock.

Feed and Fodder Resources

The major feed resources in the country are grasses and agricultural crop residues, cultivated fodders, edible weeds from cropped land and leafy fodder from trees. The quantity of fodder obtained through grazing varies from region to region and on the extent of the forest areas. The crop residues constitute major source of fodder for cattle & buffaloes. These are non-maintenance roughages and need supplementations with greens and/or concentrate for maintenance/ production.

Feed Constraints in Cattle & Buffalo Feeding

(i) Shortage of green fodder, dry fodder and concentrate in almost all the regions, leading to underfeeding of animals, and resulting in poor performance in growth, meat and milk production.

(ii) Lack of macro-and micro-elements profile in soil-plant and animal relationship in different agro climatic regions, resulted in improper or no feeding of minerals, which affects production and reproduction, leading to late maturity, longer inter- calving period and high mortality. Further, there is insufficient information about the areas rich in toxic elements (Fluorine, Selenium, etc.)

(iii) Non-application of nutrient requirements for different physiological functions of the dairy animals with the farmers.

(iv) Poor feeding management practices still being followed by the farmers, for example, not providing colostrum (to newly born buffalo calves), salt and minerals. In general proper feeding of lactating animals only is being practiced.

(v) Lack of application of low cost/no cost feeding technologies generated in the Research Institutions with the farmer's animal (UTS, UMMB, alkali treatment of straws, etc).

(vi) Absence of a mechanism of extension in feeding management.

In spite of the above constraints, the increased milk production has been possible due to slow but steady improvement in the availability of improved germplasm made available through scientific breeding, prudent use of available feed resources by the farmers, improved animal health coverage and better marketing of livestock, livestock products and by-products (casing, bones, hides and skin, etc.). Furthermore, during last 50 years, the following animal feeding technologies were also made commercially available for application in feeding the animals

- Compound livestock and poultry feeds.
- Mineral/vitamin supplements.
- Biotechnological tools like BST, pro-biotics, mould inhibitors, antimicrobials, lactobacilli, yeast culture etc.
- Synthetic amino acids like lysine and methionine.

Nutrient Requirement and Quality Fodder Availability

Rearing of large number of ruminants for milk production in India has been linked to green house effect of feeding crop residues, which increases methane production (see Chapter 14) affecting the ozone layer. A basic rule for feeding dairy cattle efficiently and to overcome the increased methane production, is to encourage cattle and buffaloes to consume large quantities of high quality roughages. They will not eat too much. In fact, the main problem is that high producing cows simply cannot digest and metabolise the necessary feed nutrients needed to sustain a heavy milk flow from roughages alone. Concentrates must be fed to the good producers to fill the gap between the nutrients required by the cow and those furnished by the roughages. Therefore, the kind and quality of roughage that is fed will dictate the type of grain mixture needed. The genetic ability of the cow to produce milk and the amount of roughage that is fed will determine the amount of grain that must be fed to milking cows and buffaloes. One of the important indications of the various studies is that the lactating cows may have higher maintenance requirement, i.e. 26.48 percent over non-lactating cows.

The higher feed intake on more transformation of digested organic matter in intermediary metabolism by lactating than non-lactating cows is responsible for higher maintenance energy requirement for lactating cows. The good dairy cows which have higher basal metabolism during lactation will need more energy for maintenance than dry cows. Similar is the case with buffaloes. Thus, a provision for energy for maintenance in dairy cattle & buffaloes should be liberal.

Protein and energy producing high quality fodders like *berseem* can produce a kg of digestible protein. A kg of digestible protein available on the other hand, if fed through concentrates will be much more costly. High quality cattle and high nutrient fodder provides the answer to sustainable feeding system for making dairying profitable. The only feasible way, which is the most economical also, of meeting the shortfall of DCP and TDN is to grow nutritious green fodder like berseem, maize, etc. It is calculated that 3.8 million hectares of irrigated land would be necessary to meet these requirements. It has been estimated that by earmarking only 4 million hectares of irrigated land to fodder production out of about 139 million hectares of available irrigated area, when backed up with inputs such as bunding, soil conservation, improved seeds and fertilizers can secure the target of sustainable fodder needs based on nutritional considerations.

Limiting Factors in Cattle & Buffalo Feeding

The major limiting factor of production in cattle & buffalo fed high roughage rations is voluntary intake. After the first meal, a cow can eat more forage only to the extent she can digest the forage already eaten. Therefore, the best indication of forage quality is the amount of forage she can eat in a 24 hours period per unit of body size. This takes into account percent digestibility, rate of digestion, size of the animal and any influence on palatability may have on voluntary intake. The factor which influences voluntary intake, and thus forage quality, to the greatest extent is stage of maturity when harvested either as green fodder, pasture, hay or silage. As forage crops mature, the plant becomes more and more lignified making the cellulose (energy) inaccessible to the rumen bacteria; this slows down rate of digestion and reduces voluntary intake. NRC in 2001 has recommended the optimum NDF concentration in the diet of cows. The effect of launching predominantly forage feeding regimes (by measured exclusion of high priced concentrate mixture) to milking herd of cattle at the National Dairy Research Institute, Karnal on milk production is portrayed in the data presented in Table 13.

It is evident from the data that on an average, the crossbred cows were capable of producing 11 to 13 kg of milk per day every month of the majority months of the year, had they been fed year round at a level of nutrient intake equal to that provided during March (when forage quality was at its peak). That high temperature is not the main limiting factor on production, which is indicated by the fact that the months lowest in average yield of milk do not coincide with the hottest weather (May-June). In order to encourage harvesting of all forage at the proper stage of maturity, which should permit the same levels of production year around that is now obtained only in March, it is suggested that the Farm Manager be motivated to accomplish this by changing his measure of accomplishment from tons of forage per year to a forage of such quality as could maintain the same or nearly same level of milk production as could be obtained in March with a minimum of concentrate feeding.

Table 13: Milk Production Round the Year based on Predominantly Forage Feeding Regime.

	All cows		Brown Swiss x Sahiwal	
	Milk (av. kg/day)	Cows (No. in milk)	Milk (av. kg/day)	Cows (No. in milk)
December	8.9	403	10.7	155
January	9.4	393	11.9	147
February	10.6	410	12.8	157
March	11.4	429	13.3	172
April	10.9	438	12.6	175
May	10.0	453	11.1	180
June	9.3	457	10.4	182
July	8.9	469	9.9	190
August	7.6	474	8.5	198
September	7.5	472	8.0	196
October	7.6	453	8.5	189
November	7.7	436	8.4	165

In an all roughage diet, that is, on a ration composed of succulent forages and small quantities of dry feeds, the total intake of feed may not be sufficient to supply all the energy required. The intake of dry matter can be increased with forages like *berseem* or *lucerne* by wilting to a certain extent which will bring down the moisture content from 80 to 90 per cent to about 55 to 60 per cent.

It nevertheless has to be borne in mind that the voluntary intake from an all fodder ration may not be sufficient to meet the total requirement of digestible energy which has to be met through supplementation of high energy concentrate mixture. As the feed intake increases, the total digestibility of the nutrients in the ration diminishes. So the intake of digestible nutrients as calculated from tables of nutritive value of feeding stuffs will be over valued. In other words, increasing feed intake (especially in high producing animals) a much larger supply of digestible nutrients will have to be provided in the form of extra concentrate mixtures. In high roughage feeding system, a lowering in the conception rate may be a problem. On excessive forage feeding, negative energy balance becomes prolonged. The results of an investigation carried out in Lousiana (U.S.A.) showed that cows losing weight had considerably lower conception rate (43 percent as compared to 65 percent in cows gaining weight). However, it is difficult to interpret this and other similar studies because too often what appears to be a simple energy deficiency is complicated by deficiencies in several other major nutrients. This subject needs much further work but it does provide a caution signal for those who might go too far with the high forage diets in high milk yielding cattles & buffaloes.

Crop Residues as an Important Feed Resource

India's remarkable progress in milk production has been attributed to multiplicity of reasons. However, two lesser known reasons are: a large increase in bovine numbers and a greater availability of crop residues. With the constant increase in cereal production in the country, there has been a matching or even greater increase in the availability of cereal straws (straw: grain ratio varying from 1:1 to 1:3 in different crops). It has been estimated that 457.7 MMT of cereal straws was available by 2000 A.D. (Anon, 2002). Thus, behind India's white revolution, perhaps its green revolution has provided a subtle support, as a spin-off benefit. Table 14 shows that crop residues contribute significantly to the total available feed energy pool. The availability of straw is not quite uniform in the different regions of the country. There are deficit and surplus areas. In northern parts of India, where large farmers use combine harvester, the paddy straw is burnt right in the field itself, rather than using it as animal feed. Apart from the facts, that the farmer wants to save labour and thinks that by burning the straws he is preparing the field for the next crop.

In the southern states, the farmers have a preference for feeding wheat straw to their animals, while in Gujarat and West Bengal farmers have preference for paddy straw as animal feed. The importance of crop residues as animal feed can be gauged by the fact that while the grain price has registered a four fold increase during the last 2 decades, the straw price has registered a 10 fold increase in some parts of the country, resulting in a fall in grain: straw price ratio from around 8 to less than 4 (Kelly and Rao, 1995). During the scarcity period, the straw is transported from surplus to deficit areas and sold at a much higher price just to save animals from starvation deaths. Even the farmers with surplus straw make a good business by selling it at an exorbitant price during the scarcity period. Apart from regional differences, the price variation could also fluctuate due to seasonal/climatic changes or the straw type. The value of straw yield per hectare can represent between 15 to 25 per cent of the total crop value. Sorghum and pearl millet straw continue to be important feed resources in western and southern parts of India as reflected by their economic values. These two being the coarser grain crops, the acreage under these crops is now decreasing. However, the value attached to sorghum and pearl millet straw by milk producers in rural and urban areas has helped to slow down the trend of decreasing their acreage in India (Kelly and Rao, 1995).

Table 14: Comparative ME availability from different feed resources for ruminants in India

Land and Type of Feed	Metabolizable energy (109 MJ) in year		
	1970	1980	2000
Permanent pastures meadows	62.8	100.5	130.0
Non-agricultural land forage	62.8	49.7	41.8
Arable land			
• Forage	1736.4	1825.0	1822.8
• Crop residues	1129.7	1330.8	1464.4
• Grain	12.6	30.1	62.8
• Oilseeds	20.9	20.9	20.9
Agro-industrial by-products	125.5	213.4	274.0
Total	3150.7	3570.4	3874.8

A survey carried out in Rajasthan, Maharashtra and Andhra Pradesh indicated that crop residues contribute 20 to 80 per cent of the dry matter available to large ruminants in different seasons. As far as the potential of crop residues, a grain yield of 3000 kg wheat per hectare produces approximately 4000 kg straw, and if a 300 kg animal consumes approximately 5 kg of straw dry matter (1.6% of body weight), then a hectare area would provide enough feed for 800 animal per days in one harvest. In spite of the high proportion of concentrate feeding practiced by urban dairies for their high yielding animals, especially in southern areas, they are forced to feed their animals enough straw to provide sufficient fibre in their diet, in order to save them from the problem of acidosis. Again in northern India where *berseem* is available in plenty during winter season, feeding of straw is essential, in order to avoid bloat condition and also to provide extra heat to the animal generated in rumen from the fermentation of straw as a protection against cold weather. Feeding of straw as fibre also helps in maintaining fat percentage in milk (Walli, 1999).

Improving Nutritive Values of Crop Residues

Nutritive value of crop residues can be improved either by way of urea treatment or by supplementing them for deficient nutrients through urea, molasses and minerals. By urea treatment of straw, energy available to the animals from basal diet can be improved by 10 to 15 percent which can help improving the milk production. At the time of storage, crop residues can be treated with 4 percent urea solution and incubated for four weeks period, using technology developed by G.B. Pant University of Agriculture and Technology, Pantnagar. Some milk sheds in the eastern and southern parts of India have successfully used this technology even with the small farmers where paddy straw is available in plenty by augmenting its nutritive value.

Urea molasses mineral block

There are various ways to supplement deficient nutrients to the animals through urea, molasses and minerals. However, urea molasses mineral block by cold process developed by the National Dairy Development Board, Anand (Gujrat) can be used for supplementing crop residue based basal diet of animals. One block of 3 kg lasts for 5 to 6 days. At a very nominal feeding cost, animals get fermentable nitrogen, energy and mineral, which helps improving the productivity of animals.

Urea molasses mineral block formulation by cold process contains

Molasses	:	45%
Urea	:	10%
Trace minerals	:	0.75%
Sodium bentonite	:	8.25%
Common salt	:	4.0%

Groundnut extraction	:	10%
RPF	:	14%
Calcium oxide	:	4.0%
Phosphoric acid	:	2.0%
Guar gum	:	2.0%

It is important to mention here that the technology for field use should be propagated in the light of the prevailing feeding and management practices.

Complete feed block

Feed resources, such as crop residues, grasses, forest by-products, fruits (jack fruit), tree leaves and several other resources are wasted in the country. If these resources are collected, processed, enriched with deficient nutrients and densified in the form of blocks of 10 or 15 kg, large amount of feed blocks can be made available for coastal areas, hilly areas or for the areas suffering from natural calamities such as drought, flood, earthquake, etc. It is heartening to note that this concept has now begun to get its footing in the country. There are few commercial companies in the country, manufacturing complete feed blocks. Straw based pellets are now being transported from Gujarat to some parts of Kerala during the last decade as wheat straw is available in Gujarat at a throw away price. Chemical composition of some of these feed pellets is given in Table 15 & 16, which can support 4 to 5 litres of milk along with maintenance, if given 8 to 10 kg per day (Gill and Garg, 1998). It is estimated that approx, 6,00,000 metric tons of jack fruit is wasted each year in North-Eastern States of India. There are reports about the use of jack fruit for feeding livestock. Ripe jack fruit, along with some urea, locally available grasses, molasses, minerals, etc. can be converted into complete feed blocks for feeding cattle & buffaloes of that region. There are lots of such examples in other parts of India where fruit production is concentrated in certain areas like apple, pineapple, etc. In these areas efforts jointly by the researchers and extension workers needs to be made for popularising its use by developing simpler technology and processes with effective extension networking.

Table 15: Enriched Wheat Straw Pellets

Ingredient	Percent
Wheat Straw	75
Molasses	15
Rapeseed cake (extractions)	5
Rice poHsh	3
Urea	0.5
Mineral mixture	0.5
Common salt	0.5
Calcite	0.5
Total	100

Strategies for Sustainable Production

Nutritional strategy

These include *inter alia* large scale and intensive use of agro-industrial by-products, development of complete feeds, strategic supplementation, wider use of leguminous forages and use of urea molasses licks.

Table 16: Production (up to 5 litres per day) and maintenance rations with enriched wheat straw

Ingredient	Per cent	
	Production ration	Maintenance ration
Wheat straw	40	60
Molasses	15	15
Rapeseed cake (extractions)	15	5.0
Deoiled rice bran	15	12
Rice polish	13	6.0
Urea	0.5	0.5
Mineral mixture	0.5	0.5
Common salt	0.5	0.5
Calcite	0.5	0.5
Total	100	100

More intensive use of agro-industrial by-products

On-farm Research (OFAR) is probably the only accurate assessment of whether new technology packages are acceptable both economically and socially to the farmers as they take into account all the interacting components unique to small farm system.

Development of complete rations

The main advantage of the system is that it (i) enables various feed ingredients to be combined through processing, (ii) extends the value of low quality roughages, and (iii) reduces feed wastages and feed costs. The development and use of complete diets merits increased attention in the future.

Strategic supplementation

Strategic supplementation of energy, protein, minerals and vitamins is essential, especially when availability is limited. Supplementation should be considered in the context of seasonal requirements, physiological state of the animal and for different production systems.

Use of leguminous forages

In India, berseem *(Trifolium alexandrinum)* and throughout South-East Asia, cassava leaves and lucaena are very widely used to feed buffaloes and cattle. Their ISPA use can be now widely applied because of several advantages. These include *inter alia* (i) availability in the farm, (ii) accessibility, (iii) provision of variety in the diet, (iv) source of dietary nitrogen (N), energy, mineral and vitamins, (v) laxative influences on the alimentary system, (vi) reduction in the requirements for purchased concentrates, and (vii) reduced cost of feeding.

Urea molasses block licks

These blocks make available energy, non-protein nitrogen, minerals, and vitamins and possibly also medication to the animals which are fed on crop residues.

Transferable Feeding Management Interventions

Feeding management is critical in crossbred high yielding cows if the dry matter intakes/nutrient intake of very high magnitude is going to be attained and sustained for sustaining milk-yielding potentials.

Following feeding management interventions are suggested for sustainable dairy farming:

1) Complete feed block

Concept of feeding complete feed blocks may be popularizing in feed/fodder scarcity areas/periurban dairies, areas where labour cost and storage cost is high and also among farmers having poor cattle & buffalo feeding knowledge.

2) Oestrus synchronization with fodder availability

Oestrus synchronization with fodder availability to get more milk and to reduce cost of milk production should be followed. Producing milk on *berseem* fodder during winter month not only increases production but reduces feeding cost; improve animal health in the northern states of India.

3) Grouping of cows and buffaloes

Grouping cows with ample space for feeding is required to reduce negative social interaction and to minimizing feeding competition. Segregating heifers during the transition period and during their first lactation is good for better production. Heifers have greater growth requirements, greater persistency of lactation during their first lactation. On the contrary, due to smaller body size and frequently low position in a group dominance hierarchy. They must be segregated from pre calving through their entire lactation.

4) Use of bypass protein

The bypass protein feeding has definite advantage in improving milk production potential. On an average on feeding 1kg bypass protein supplement daily milk yield, fat and protein percent increase by 0.5-1.0 kg, 0.2-0.5% and 0.2-0.3%, respectively. Protection of protein can be achieved by treating oil cakes with formalin solution (1.2 gm formaldehyde per 100g CP for treatment of mustard and ground nut cake, after treatment cake is kept under air tight condition in polythene bags for 1-2 days) or by heat treatment.

5) Use of fat / oils

Oil is source of concentrated energy (2.25 times higher than starch) that can increase milk yield and fat test. Early lactation (120 days postpartum) when cows are in negative energy balance (losing body weight) general recommendation is 150gms/cow/day of supplemental oil. During hot humid months, fat feeding reduces the heat stress edible oils (Mustard/Till oil) @1-1.5% of concentrate mixture may be suggested for sustainable milk production when animal is under any climatic or physiological stress like weakness after sickness.

6) Feeding sodium bicarbonate

Feeding sodium bicarbonate maintains desired pH (6.2 to 6.8) in the rumen, this improves feed intake and digestibility, corrects volatile fatty acid production, and can help maintain normal milk fat test. Feeding sodium bicarbonate @1% of concentrate when off-feed is a problem, such as during the early postpartum period, or when the ration dry matter from roughages is low (below 50%). Feeding carbonated feeds (concentrates containing 1% sodium bicarbonate) are important for high yielding cows and buffaloes.

7) Feeding and caring of dry animals

Feeding and caring of dry animals will also helps in subsequent lactations thus, sustaining milk production at dairy farms. Dry cow therapy must be practised to avoid the chances of *Mastitis* in subsequent lactations.

8) Particle size of feed

Keep watch on particle size of feed offer and feed particles in dung. Feed refusals give fairly good idea to improve feed mixing. Manage and enhance energy density of ration. Provide sufficient fiber to avoid acidosis, erratic DMI, laminitis, ketosis, decreased milk and milk fat.

9) Alternative feed sources

Using alternative feed sources and supplementation providing mineral mixture, concentrate and green forages should modify the traditional feeding practice.

10) Urea source of NPN (Non Protein Nitrogen)

Urea is source of non-protein nitrogen for ruminants. One percent incorporation into the grain mix/concentrate mixture. Maximum of 100 gms of urea/cow/day in total ration may be recommended. Similar recommendations are made for buffaloes.

11) Zinc methionine

Zinc methionine a source of zinc and methionine is important ingredient in lactating animals feeding. Organic chelated form plus methionine, is not degraded in the rumen. Four and a half (4.5) gm per cow/day is suggested for high producing cattle & buffaloes with feet and leg problems.

12) Chemical preservation of fodder

Green fodder (non-leguminous e.g. maize, jowar, oat, barley etc.), should be harvested at around 20 to 25% dry matter. The fodder must be chaffed in a chaff cutter. One quintal lots of this fodder should be made and on each lot add 500 g urea plus 1 kg common salt and mix it thoroughly in the fodder. Fill the treated material tightly in an earthen pit or a concrete pit and compress so as to accommodate more material. Then cover it with dry material like maize/bazra/jowar stroves followed by a polythene sheet. After a period of 2-6 months or at the time of necessity the material can be gradually taken out and fed to the animals. The preserved fodder when fed, improve the milk production during fodder scarcity period, which leads to sustainable milk production.

13) Urea (Ammonia) treatment of straw

The technique is more suited to Indian farmers having low to medium milk producing cows/buffaloes; the straw is treated with 4% urea at 40% moisture level and ensiled for 20 days in stacks. Thus, for 100 kg of wheat or paddy straws 4 kg of urea is required to be dissolved in 50 litres of water. This urea solution is then mixed with the straws and the material is stacked in the form of bonga or under polythene cover. After a period of 3-4 weeks, the stack can be opened and the material can be fed to the animals directly without any addition of molasses but mineral mixture and a source of vitamin A must be added to meet the essential requirements of the animals. This technique will be useful when lot of straw is used in the ration.

14) Location specific mineral mixture formulation

It is important to use Location Specific Mineral Mixture Formulation in dairy animal feeding for sustaining milk production (Table 17). Further significant interaction occurs between trace elements and immune function. Of particular importance are deficiencies of trace elements that can impair immune function before clinical sign of nutrient deficiencies are evident. Additionally some effects of trace element deficiencies on immune function may be long lasting after repletion has occurred.

Table17: A Mineral Mixture Formula (20-Kg Batch) Developed At NDRI Karnal For Areas Around Karnal (Haryana), India is Given Below:

Dicalcium phosphate	8.2 kg
Calcium carbonates	2.5 kg
Magnesium sulphate / carbonate	0.3 kg
Common salt	6.1 kg
Trace mixture	
Copper sulphate	73 gm
Mangnous oxide / Mag. dioxide	243 gm
Cobalt chloride	1.5 gm
Potassium iodide	4.0 gm
Sodium fluoride	1.0 gm
Zinc sulphate	655 gm
Ferrous sulphate	922 gm

15) Suitable feeding schedule

Suitable feeding schedule has important role in sustaining milk production. Following Nutrient Requirement (kg/day) for planing feeding schedule may be considered for feeding dairy cows to sustain their milk production (Table 18)

Table18: Feeding Schedule Based on Milk Production

SN	Category	Nutrient Requirement (kg/day)		
		DM	CP	TDN
1.	Yielding upto 5.0 kg	13.50	0.92	5.50
2.	Yielding upto 5-8 kg	13.50	1.20	6.60
3.	Yielding upto 8-11 kg	13.50	1.49	7.70
4.	Yielding upto 11-14 kg	16.00	1.75	8.80
5.	Yielding upto 14-17 kg	16.00	2.07	9.90
6.	Yielding upto 17-20 kg	16.00	2.36	11.0
7.	Yielding upto 20-23 kg	18.00	2.64	12.1
8.	Yielding upto 23-26 kg	18.00	2.89	13.2
9.	Yielding upto 26-29 kg	18.00	3.13	14.3
10.	Yielding upto 29-32 kg	18.00	3.45	15.3
11.	Yielding above 32 kg	18.00	3.77	16.3
12.	Calves from 3-6 months	2.00	0.26	1.40

Contd...

13.	Calves from 6-12 months	3.70	0.44	2.40
14.	Calves from 1-2 years	7.50	0.64	4.50
15.	Calves above 2 years	10.20	0.81	5.30
16.	Dry stock	11.20	0.82	4.80
17.	Down calvers	11.20	0.82	4.80
18.	Growing bulls	13.70	1.11	7.30
19.	Bullocks/Teasers	13.70	0.79	5.20
20.	Bulls	13.70	0.83	6.10

16) Feeding based on body condition

Researchers have shown that a relationship exists between production or health problems (such as dystokia, cystic ovarian disease, foot problems, metritis) and change in body condition score. Milk producer should use body condition score as an indicator of nutritional status of their herd. Body condition gives as indication of how the animal has been fed over the preceding weeks/months. The level of milk production of the dairy cow also affects it.

To assess the body condition of dairy cows objectively a procedure called body condition scoring may be followed. A simple body condition scoring scale is given below (Table 19):

Table 19: Evaluation of Dairy Animals Based on Body Condition Score

Score	Evaluation	Description of Animal
Score 1:	Very poor	Animal emaciated, deep cavity under tail head, no muscle covers between pelvis and skin all the bones very prominent.
Score 2:	Poor	Animal appears weak, cavity under tail head is marked, and deep depression is loin area, some muscle mass evident all over the body.
Score 3:	Moderate	Shallow cavity in tail head region seen, slight fatty tissue also evident, pelvis can be felt easily, depression in loin region still evident. Ends of transverse processes of lumber region can be palpated with some pressure.
Score 4:	Good	Fatty tissue can be felt all over the animal body (chine, loin and rump region). Skin appears smooth but pelvis can be felt. Ends of transverse processes can be felt but thick layer of tissue in the region evident. Only a slight depression is evident in loin area.

Contd...

Score 5:	Fat	Folds of fatty tissue present all over, pelvis felt only with firm pressure. Transverse process is difficult to palpate. No depression in the loin is visible.
Score 6:	Very fat	Tail head buried in fat tissue, skin distended, Pelvis can not be felt even with firm pressure, fat accumulation over transverse processes evidence. Bony structure is not palpable.

Recommendations

Monitoring condition score over the lactation cycle is very useful in focusing on management and nutritional areas for further improvement. Optimum body condition the maximum milk production may be higher than that for optimum milk production (4-4.5 v/s 3-3.5). Feed conversion efficiency of a fat animal is lower. Excessive loss of body condition (> 1.0 point) after calving is more harmful than higher condition itself. The owner must ensure that the cow or buffalo does not lose body weight excessively after parturition. This has adverse effect on milk production, reproduction and health of the animal. The body condition of the cow should not go below 2.5 - 3.0 even during peak lactation. This should be ensured in high yielder by: maintaining the optimum body condition score at calving would be around 4-4.5. Animals fattier than this at calving are likely to develop fatty liver syndrome during early postpartum and may not sustain her milk yielding potential.

17) Increasing light hours for feeding

Increasing light hours for feeding by artificial lighting helps in sustaining milk production in high yielding animals. Tests show that by using artificial light to extend daylight to 16-18 hours during autumn and winter months. the milk yield can be increased by 10-12%. Illumination strength recommended is between 100-300 lux.

18) Feeding & milking during cooler hours

Feeding & milking during coolers of the day will further improve the milk production.

19) Feeding cold water in summer months

Feeding cold water in summer months improve water intake, reduces heat stress and help in sustaining milk production potential of high yielding dairy animals.

Summary

For sustainable feeding practices, judicious use of home grown fodders and concentrates is essential. Improved forage quality from harvesting at the proper stage-Goal: 60-65% TDN (NRC) for forages other than maize silage. Greater green forage consumption-Goal: 40-45 kg per cow. Higher protein forage from a greater legume content-Goal: minimum of 13-14% on dry matter basis. Utilizing strategies developed so far for sustainable milk production in cattle and buffaloes. Reduced grain intake, goal should be one kg less per day. Feed wastage should be minimised with better storage, handling and feeding practices as well as mould prevention.

□□□

Chapter 14

Methane Emission from Cattle & Buffalo

Introduction

Sustainability comprises a component of "prevention of environment degradation" so that the natural resources are used efficiently in present and at the same these available resources should be saved for future use. Ruminants are often blamed for excessive methane emission. Methane gas is one of the major green house gases and has contributed in global warming / green house effect. Methane emission in cattle begins approximately four weeks after birth, when solid feeds are retained in the reticulorumen (Anderson *et al.*, 1987). Fermentation and methane production rates rise rapidly during reticulorumen development. Estimates of yearly methane production of the typical beef and dairy cow range from 60 to 71 kg and 109 to 126 kg, respectively (EPA, 1993). Measurements made by indirect respiration calorimetry show methane losses vary from approximately 2 to nearly 12% of GE intake (Johnson *et al.*, 1993b). Generally, as diet digestibility increases, variability in methane loss also increases. The methane gas emission cannot be stopped from ruminants, thus the viable option is to focus on decreasing the methane gas production from the ruminants. Many factors influence methane emissions from cattle and include the following: level of feed intake, type of carbohydrate in the diet, feed processing, addition of lipids or ionophores to the diet, and alterations in the ruminal microflora. Manipulation of these factors can reduce methane emissions from cattle and buffaloes. Methane production in buffaloes is comparatively lesser than the cattle. In India, buffalo is the major milk animal hence, the contribution to green house gas can be decreased by encouraging buffalo breeding. We need to adopt the selective breeding approach with respect to methane gas emission in cattle and buffaloes. But the breakthrough can be achieved via altering the feeding regime and by giving attention to nutritional strategy in dairy animals, as described in chapter 13.

Mechanisms of Methane Production in Ruminants

There are two chief mechanisms that cause variation in methane production from cattle and buffaloes.

The first is the amount of dietary carbohydrate fermented in the reticulorumen. This method has many diet-animal interactions that affect the equilibrium between the rates of carbohydrate fermentation and passage.

A second mechanism regulates the existing hydrogen supply and subsequent methane production through the ratio of volatile fatty acids produced. Predominantly, the fraction of propionic acid that is produced relative to acetic acid has a major impact on methane production. If the acetic:propionic acid is half (1/2), the loss of substrate energy as methane would be zero (0%). If all carbohydrate is fermented to acetic acid and no propionic acid is produced, energy loss as methane would be 33% (Wolin and Miller, 1988). Because acetic:propionic acid typically varies from approximately 0.9 to 4.0, corresponding methane losses vary widely as well. Methane production can also be affected when there are significant alternative hydrogen sinks are present. These unusual sinks are usually relatively minor but can include oxygen, unsaturated fatty acids, nitrates, sulphates, and microbial growth.

Methane Production from India

India possesses the world's largest livestock population with a high degree of diversity in its composition. Among the various livestock species, cattle dominate with 38.2% of the total methane produced in the country followed by goat (25.7%), buffalo (20.2%), sheep (12.7%)

and others (3.3%) (Abha Chhabra *et al.*, 2009). The detailed state/district-level methane emission inventory by age-groups, indigenous and exotic breeds of different livestock categories was estimated using the country-specific and Indian feed standard based emission coefficients and recent livestock census 2003. The total methane emission from Indian livestock, which includes enteric fermentation and manure management, was 11.75 Tg for the year 2003. According to an estimate by Abha Chhabra *et al.* (2009) enteric fermentation accounts for 10.65 Tg (~91%) compared to 1.09 Tg (~9%) by manure management. Dairy buffalo and indigenous dairy cattle together contribute 60% of the total methane emission. The three high methane emitter states are Uttar Pradesh (14.9%), Rajasthan (9.1%) and Madhya Pradesh (8.5%). The detailed district-level spatial analysis in GIS environment resulted in the identification of clusters of districts with high emissions. Among these, Mednipur District (West Bengal) reported the highest total methane emission of 0.12 Tg. Using the remote sensing-derived livestock available feed/fodder area, the average methane flux from Indian livestock was computed as 74.4 kg/ha.

Factors Effecting the Methane Production

Many factors have their effects through one or more of these mechanisms. These factors include feed intake, type of carbohydrate, forage processing, lipid addition, and manipulation of ruminal microflora including the use of ionophores.

Effect of amount of feed intake

Feed consumed by any given animal if increases, the percentage of dietary GE lost as methane decreases by an average of 1.6% per level of intake (Johnson *et al.*, 1993b). Efforts to use statistical relationships to calculate this decline in methane production with increased intake have largely failed (Blaxter and Clapperton, 1965) and limit extrapolations from laboratory to field situations. When highly available carbohydrates are fed at limited intakes, high fractional methane losses occur. At high intakes of highly digestible diets, low fractional methane losses occur.

Different type of carbohydrates and their effect on methane production

The type of carbohydrate fermented influences methane production most likely through impacts on ruminal pH and the microbial population. Fermentation of cell wall fiber yields higher acetic:propionic acid and higher methane losses (Moe and Tyrrell, 1979; Beever *et al.*, 1989). Moe and Tyrrell (1979) found fermentation of soluble carbohydrate to be less methanogenic than cell wall carbohydrates. As a greater amount of any carbohydrate fraction is fermented per day, whether it is fiber or starch, methane production is decreased. Methane losses with high intakes of beet pulp fell to 4 to 5% of GE intake (Kujawa, 1994). Additionally, the fermentation of brewery and distillery products containing relatively available fiber results in a surprisingly low methane production, generally one-half to one-third of that seen with common feedstuffs of comparable digestibility (Wainman *et al.*, 1984). The very high grain diets (plus 90% concentrate) commonly fed in U.S. feedlots result in strikingly different methane loss rates than are commonly predicted. Considerable variation is found among diets, but typical losses frequently fall between 2 to 3% of GE (Abo-Omar, 1989; Carmean, 1991; Hutcheson, 1994). This loss rate is approximately one-half of the commonly predicted 6% of diet GE lost as methane.

Forage processing

Grinding and pelleting of forages can markedly decrease methane production (Blaxter, 1989). These effects are not apparent when intakes of these diets are restricted, however. At high intakes, methane loss/unit of diet can be reduced 20 to 40%. Increased rate of passage of the ground or pelleted forage likely contributes to the reduced methane production. Okine *et al.* (1989) reported a significant reduction in methane production (29%) when weights were added to the rumen. Methane production was reduced from 189 to 135 L/d with no change in digestibility.

Addition of lipids

Fat additions to ruminant diets impact methane losses by several mechanisms, including biohydrogenation of unsaturated fatty acids, enhanced propionic acid production, and protozoal inhibition. Czerkawski *et al.* (1966) demonstrated that addition of long-chain polyunsaturated fatty acids decreased methanogenesis by providing an alternative metabolic hydrogen acceptor to reduction of CO_2. However, the amount of total metabolic hydrogen used in the biohydrogenation process of endogenous unsaturated fatty acids is small (1%) compared with that used for reduction of CO_2 to methane

(48), VFA synthesis (33%), and bacterial cell synthesis (12%); (Czerkawski, 1986). Sheep or cattle fed supplemental fat sources such as animal tallow or soybean oil had decreased methane production compared with controls fed isocaloric diets (Swift *et al.*, 1948; Haaland, 1978; Van der Honing *et al.*, 1981). The reduction in methane production in these studies was attributed to decreased fermentable substrate rather than to a direct effect on methanogenesis. The methane reduction effects are likely to be dramatic only when basal digestion is inhibited, such as the 29% reduction in methane found by Park *et al.* (1994). In this experiment, the addition of 6% of a dry powdered fatty-acid supplement to a 60% concentrate diet also marginally decreased organic matter digestion.

Addition of ionophore

Ionophore additions to beef cattle diets, particularly monensin, reduces feed intake 5 to 6%, decreases acetic:propionic acid and decreases methane losses (Goodrich *et al.*, 1984). However, recent investigations indicate that the decrease in methane production is short-lived (Rumpler *et al.*, 1986; Abo-Omar, 1989; Carmean, 1991; Saa *et al.*, 1993). Methane production per unit of diet by cattle fed either grain or forage diets returned to initial levels within 2 week. The reduction seen in methane production by ionophore supplemented cattle is possible to be related to the reduction in feed intake and not a direct effect on methanogenesis.

Microbial flora alterations

Ruminal protozoa may also play an important role in methane production, particularly when cattle and buffaloes are fed high-concentrate diets. Ruminal methanogens have been observed attached to protozoal species suggesting possible interspecies hydrogen transfer (Stumm *et al.*, 1982). Defaunation of the rumen of cattle fed a barley diet decreased methane production by approximately one half (Whitelaw *et al.*, 1984). However, defaunation of animals receiving high-forage diets (Itabashi *et al.*, 1984) did not reduce methane losses.

Summary

The development of management strategies to alleviate methane emissions from cattle is possible and desirable. Strategies in reduction of methane gas emission will enhance utilization of dietary carbon, improve feed efficiency and animal productivity. A decrease in methane emissions will reduce the contribution from ruminants, particularly cattle & buffaloes to the global methane account. If the ruminants of a particular country have a lower methane emission and higher farm production this creates a positive impact on sustainable animal husbandry practices.

□□□

Chapter 15

Housing and Heat Stress Management for Cattle & Buffaloes

Introduction

Sustainable animal production is a system, which preserves the basis of life of future generations. It involves design and management procedures that work with natural processes to conserve all resources, minimize waste and environmental damage, while maintaining or improving farm profitability. Sustainable animal production are designed to take maximize advantage of existing animal production as well. Such systems have tended to avoid the use of synthetically compounded growth regulators and livestock feed additives etc. and aims to produce food that is nutritious, without being contaminated. Animal production in developing countries like India utilizes the full range of animal outputs, many of which are returned as essential inputs to the farm production system. Livestock serves as a faster means for recycling nutrients and as a source of energy and value-added production. Their complementary role within the farming system is unique and needs to be fully exploited. As in conventional animal production system, the success of sustainable approaches is very dependent on the skills and attitudes of the producers. The degree to which different models of such farms are sustainable is very variable, and is dependent on the physical resources of the farmer, and the degree deficiencies in support farm, the talents and commitment of the support available.

Housing is the most important part of the animal management system. Animals require protection from high and low temperature, strong sunlight, heavy rainfall, high humidity, frost, snowfall, strong winds, ectoparasites and endoparasites. Climatic stress occurs when the temperature goes beyond the critical limit. High humidity combined with high temperature causes more stress (Temperature Humidity Index - THI). Strong winds further aggravate the conditions. Good housing provides hygienic conditions, better utilization of feed and labour, proper supervision, clean milk production and proper recycling of by-products.

The country has been divided into 15 agro-climatic regions based on similarities in cropping conditions like rainfall, temperature, soil, topography, cropping and farming systems, water resources, etc. Though agriculture is considered to include animal husbandry, there is practically no classification of zones for animal husbandry in the country. However, closely interlinked livestock production with crop farming in a given region/zone is mentioned. It has its own grounds based on which it prospers or withers.

Different Animal Husbandry Regions

India can be divided into the following five animal husbandry regions based on rainfall, type of soil, topography and temperature of the region.

1) Temperate Himalayan region

This region comprises of the Himachal Pradesh and Jammu and Kashmir, Uttarakhand, hilly regions of Punjab and West Bengal, Sikkim, Assam and other north eastern states severe soil erosion in the hilly regions in the main region are the main problem in this region. The region has high rainfall and thick forest cover. Livestock population of these areas are largely raised on pastures. During severe cold months, animals are moved from higher altitude to the villages in the foot hills.

Basic principles of the housing in these areas are 1) Housing should be sloppy so that the excess water may not be accumulated. 2) Provision of bedding especially during winter months with straw and provide gunny bag to protect the animals from chilly winds.

2) Dry Northern region

This region comprises the plains i.e. Punjab, Haryana, Delhi, Western Uttar Pradesh and part of Gujarat and Madhya Pradesh. Annual rainfall in this region is medium to low. Due to the favourable environment for livestock several important breeds of cattle and buffaloes are found in this region.

Basic principles of housing : Loose housing system is most suitable in this region.

3) North-Eastern region

Plains of West Bengal, Bihar, Orrisa, eastern part of Uttar Pradesh, Assam and Andhra Pradesh comprises of this region. The annual rainfall in these regions varies from medium to high. Most of the non-descript breeds are found in this region. Productivity of the animal in this region is low due to lower availability of good quality feed and fodders. Animals are depending on the straw and other crop residues. Large animals are used for the agriculture operations and carting.

Basic principles of housing in this region are 1) Roof should have gentle slope 2) Adequate ventilation facilities.

4) Southern region

This region comprises the part of Madhya Pradesh, Maharashtra, Andhra Pradesh, Tamil Nadu and Karnataka. Rainfall is generally low and the atmosphere is quite humid. This area is the home tract of some of the best draft breeds of cattle.

Basic principles of housing in this region are 1) Close housing and 2) Proper ventilation facilities.

5) Coastal region

This region comprises coastal belts of Maharashtra, Karnataka, Kerala, Tamil Nadu, Andhra Pradesh and Orrisa. Annual rainfall in this region is very high (>500cm). Most of the time, this region remains humid due to high rainfall. Animal husbandry in this region is least developed and animals live on poor quality paddy straw.

Basic principles of housing in this region are 1) Adequate arrangement to protect the animals from the heavy rainfall 2) Roof should have gentle slope and 3) Adequate ventilation facilities to get rid from the high humidity.

Considerations for Housing

The animal house should be conveniently located for profitable business. Market for sale of milk should be available nearby at minimum cost on transportation. Availability of water and electricity are essential for operation of dairy farms. The land at which the housing unit is proposed should not have water logging problem. The not-so-well drained area can cause out break of water borne diseases. Wind direction in the area should be observed for the purpose of locating doors and windows. The farm building should be located at a place, which has maximum sun exposure as a part of health cover. Shady trees should be planted on the boundaries of the building so the animal can be protected against the chilly wind and provide shade during hot summer months.

Housing Systems

Dairy animals are housed in shelters and yards. Shelter is a shed, which has roof resting on pillars or walls. Yard is a fenced area with a gate. Shelter along with yard; give rise to three type of housing system as follows.

A. Confined housing

In this type the animals are kept in close confinement and tied at all the times. The milking, feeding and other operations are carried out at the same place. The barn is completely roofed and walls are also complete with windows and ventilations located at suitable place. It is practiced where land cost is high particularly urban area. It gives better protection from extreme cold or hot climate and high rainfall. As the animals do not get exercise they eat less and their performance is comparatively lower. They become prone to limb defects as they stand on hard floor for long hours. Heat detection is poor, as mounting behaviour cannot be expressed by confined animals.

B. Loose housing

Loose housing may be defined as a system where animals are kept loose except at the time of milking or during treatment. The main advantage of the system is that the cost of construction is comparatively low and there is flexibility of adaptation. After feeding and milking in shed, animals move freely in the yard and get exercise. Animals eat more, performance is better and heat detection is comparatively easier. The labour cost is reduced under loose housing units, the animals have more freedom of movement, and therefore they remain in a comfortable condition.

Larger land requirement, faster spread of disease among loose animals and less individual care are some of the demerits of the system.

C. Open housing

Open housing system has only yard with tree shade. No shelter is provided to protect animals from adverse climate. It is practiced in moderate zone in rural area where farmers cannot afford to invest on construction of shelter.

Space Requirement

Covered and open space required per animal depends mainly on the age, and on the size of the animal. However bulls and down-calvers (cows nearing parturition) require more space due to obvious reasons. Another important aspect to be taken care of in animal houses is the provisions of correct length and size of managers and water tanks so that all animals can have access to feed and water always. The requirement of floor space, manager or feeding space and watering space are given below (Table 20):

Table 20: Floor feeding (manger) and watering space requirement of dairy cattle under loose housing system

Type of animal	Floor area (M^2)/ animal		Manger space (cm/animal)	Watering space (cm/animal)
	Covered	Open		
Cows	3.5	7	60-75	60-75
Down calvers	12.0	12	60-75	60-75
Bullocks	13.5	7.5	60-75	60-75
Suckling calves	1.0	2	40-50	10-15
Older calves	2.0	4	40-50	10-15

Note:

* Sufficient watering space has to be provided in each pen to accommodate two percent of the animals of shed simultaneously

** Under group management the watering space could be reduced depending upon the size of group as not more than 20% cows will go to water trough at point of time.

Orientation of Shed

The best orientation of shed depends on cattle & buffalo management and the type of flooring under the shade. For situations with dirty floor and cows able to move outside the structures, a North-South orientation is recommended (Figure 10). This allows sunlight to dry 50 percent or more of the floor while still providing shade outside the structure. For confinement facilities with concrete floors, an East-West orientation provides maximum roof.

Figure 10: Cattle and Buffalo Shed

Sustainable animal production meets economic, environmental, and social objectives simultaneously; since these three objectives always overlap, they are managed together. It is generally assumed that under Indian conditions, except heavy precipitating areas, loose housing system is suitable. In winter night's loose housing system is stressful to the animal because of extreme cold and chilly winds in northern parts of India so adequate arrangement is necessary to protect animals from cold wind. Wind breaking curtains provides good relief to animals. Animal shelter should alter or modify the environment for the benefit of the animals enclosed in it. Economy should receive special attention in their initial cost and maintenance, health and comfort of animals.

Heat Stress Management

Livestock production including dairying is an integral part of traditionally diversified system of Indian agriculture, which has come to stay as a complementary and supplementary component of agricultural economy. At present the contribution of livestock sector in the over all agriculture economy is on the rise.

The performance of any species including dairy cattle or buffaloes depends on the influence of heredity and environment factors. Keeping aside the genetic make up of the individual, management practices play an important role for improving the over all productive performance of animal to a great extent.

In general, environment of an animal is the condition to which it is subjected in its usual course of life. The environment during extreme summer has temperature up to 45 degree centigrade and has relative humidity of about 80%. This high temperature coupled with high humidity imparts greater stress on dairy animals and is measured as temperature and humidity index (THI). Crossbred cattle and buffaloes are more sensitive to thermal stress. Generally heat stress results in inability to dissipate sufficient heat from the animal to maintain homoeothermy this is primarily due to high air temperature but is intensified by high humidity, thermal radiation, low air movement and metabolic heat. Heat stress affects both production as well as reproduction of the dairy animals. Therefore, it is an important effective measure to mitigate these stresses.

Effects of Heat Stress

1. Decrease in growth rate

Summer stress reduces the growth rate of cattle & buffaloes particularly during their growing phase probably through reduction in feed intake and changes in hormonal physiology in the animals. Under thermal stress animal needs to divert a greater portion of its energy to maintain its body temperature by panting and sweating rather than growth. This ultimately leads to delay in maturity and prolonged age of first calving.

2. Reduction in feed intake

One of the reactions of cattle & buffaloes in elevated temperature beyond the thermal neutral zone is marked by decreased feed intake in animal. The extent of this depression is proportional to the thermal stress. Feed efficiency is also reduced which may be due to increased physiological activities involving expenditure of energy for maintaining body temperature. These increased activities result in increased heat production above an environment temperature of 30 degrees. It has been found that during warmer months feed efficiency goes down by 10-12% in lactating animals compared to cooler months.

3. Decline in milk production

Decrease in milk production in summer months is due to poor quality and quantity of forage supply. Due to this, body receives less total energy than required and milk production goes below normal.

4. Change in milk composition

High ambient temperature causes depression in yield of milk fat, SNF and casein content of milk. Lactose percentage is also reduced when animals are exposed to high ambient temperature.

5. Increase in water intake

Water consumption increase as much as 40% when air temperature rises to about 35-37 degree centigrade because due to heat stress animal losses large quantity of water by evaporation through their lung and skin.

6. Decrease in reproductive performance

Heat stress lowers the fertility in dairy animals particularly in buffaloes. Due to increased cortical synthesis, there will be blockage in action of oestrogen hormone, which leads to sub-active ovaries, silent heat and anoestrus. Some other complications also occur due to thermal stress like low conception rate, repeat breeding, early embryonic death and abortion.

Heat stressed bulls also shows poor libido and decrease in semen quality and quantity.

Measures to Counter Heat Stress

To counter the effect of heat stress three basic managemental strategies may be adopted.

A. Physical alteration of environment
B. Change in feeding regime.
C. Adaptation of proper breeding strategy

A. Physical Alteration of Environment

To alleviate thermal stress in summer, the most important point is that the cattle & buffaloes should be protected from exposure of direct sun for long period. This can be achieved by following the steps mentioned below.

1. Proper sheltering and housing

i) Animals may be housed during the hot part of the day in simple thatched roof having half walled sheds. Height of roof should be 3.5 to 4.0 meter, preferably A-shaped, because one side of the roof will save other half side from the direct solar radiation by casting its shadow which helps in cutting down heat gain from roof of the shed. It has been observed that dairy animals remain more comfortable and perform better in loose housing system than closed barn type shed. Animals should be provided sufficient housing space at least 4-5 square meter covered area and 8-10 square meter open area should be provided to each adult dairy animal as a general rule.

ii) During severe hot days accompanied by hot surface wind, the open end of the shed may be covered with the thatched bamboo/ Jute / Hessian curtain. Such practice has been found to be effective and beneficial to the animals.

iii) Roof material should be bad conductor of heat. Insulation capacity and colour of roof has great influence on protection potential of shed against direct and indirect solar radiation.

Painting the roof top with white colour reduces direct solar radiation in summer. The under surface should be painted with dark colour.

iv) Developing grass cover, shrubs or creepers around the shed can also reduce the unfavourable effect of solar radiation. A shady tree is a good and cheap way of providing protection to animals from heat radiation.

v) A fast growing tree wind breaker, located 25-45 m from building with 3-8 rows of tree around 6-10 m wide, can also prevent the hot wind during summer season.

vi) During the night the animals may be left in open. They can then get rid of most of the heat gained during the day time by radiation.

2. Water application

During summer wetting of cattle and buffaloes can effectively reduce rise in their body temperature due to evaporative cooling. This may be achieved by

i) Splashing / Shower

Splashing water on dairy animals particularly buffaloes in side the shed 3-4 times during the hottest part of the day (11 am to 3 pm) is effective. It has been found that buffaloes produce more milk when their body was wetted through splashing.

ii) Sprinkler

The most practical way to relieve heat stress is wetting the animal with low pressure sprinkler which sprinkle large droplets (Figure 11). To reduce body temperature the dairy animals can be sprinkled in the holding area before milking. Droplet size is the key factor of this cooling system enabling penetration of water through their coat for skin contact. Sprinkling buffaloes under shade during summer improve feed consumption and rate of weight gain.

iii) Wallowing

In villages buffaloes are taken to pond for wallowing. Making them walk to and from the pond in the hot sun neutralizes any beneficial effect that can be expected from wallowing. So wallowing arrangements should be made at the farm.

Figure 11: Mist cooling System to counter the Heat Stress

iv) Cooling Devices

On a large commercial dairy farm various types of cooling devices like fans, wet curtains or panels, air coolers or air conditioning can be used during hot summer condition.

B. Change in Feeding Regime

1. Feed intake

i) Simple nutritional management includes increasing the number of feeds (upto 6-8 times) per day after providing more palatable high quality and low fiber diet. Decreasing the proportion of fiber in the diet has several potential benefits during the period of heat stress. Forages generate a greater heat increment than concentrates; therefore decreased dietary fiber may decrease heat production by the animals.

ii) To maintain energy with decreasing DM intake. The energy of diet should be increased. This can be achieved by decreasing the forages is to concentrate ratio and increasing forages quantity and also by adding fat to ration. This provides greater energy during hot weather. It has been found that 3-5% fat can be added to ration without toxic effect on ruminants.

iii) Propagation of green fodder should be increased.

iv) During heat stress dietary protein density should be increased to compensate for lower feed intake.

v) Dietary mineral concentration should be increased during heat stress to compensate for decreased dry matter intake and increased excretion of certain minerals during summer months. Potassium is a major mineral secreted during sweating. Sodium excretion increases in urine during heat stress in dairy animals. Supplementing dietary sodium and potassium increases milk yield during the period of heat stress.

vi) It is advised to take the animals for grazing only in the cooler hours of the day i.e. in the early morning and late in the evening. If possible, even night grazing can be practiced. It has two folds benefit. First, during night animal will not expose to heat stress and secondly, they can then get rid off most of the gained heat during day time by radiation.

2. Water intake

Water is the most critical nutrient for all animals and its requirement is intensified during hot climatic conditions. Following steps should be adopted:

i) Water should be always made available to the animals in their close proximity.

ii) Water should be of good quality, it should be soft, clean, fresh and devoid of undesirable odour and taste.

iii) Provision of cooled drinking water also helps in reducing the heat load on dairy animals. Shaded water trough in sufficient number should be provided to the animals at convenient places in the farm.

C. Adoption of Proper Breeding Strategies

1. Selection of heat tolerant strains within breed should be adapted for future breeding.
2. Protection to breeder bulls and teaser against thermal stress should be provided. For this preferably heat detection and breeding should be carried out in cool hours (either in early morning or late in the evening) of the day.
3. Young teasers may be used for heat detection during summers due to high potentiality of young bulls.

4. Artificial Insemination in dairy animals should be scheduled during cooler hours of the day for high fertility and conception rate. It should be carried out with good quality semen at optimum time by experienced personnel only.

Summary

It is a proven fact that proper construction of animal sheds can increase the production from animals. The alleviation of heat stress among cattles, crossbreds and buffaloes will lead to sustainable milk production round the year. The temperament and milk yielding capacity of the dairy animal can be improved with better housing facilities. Moreover, it our moral duty under animal welfare to provide the better living conditions. Lowering the stress caused by extreme winter and summer season in animals is a mandatory step in scientific dairy management.

❑❑❑

Part-3

Impact Assessment of Sustainable Breeding

Chapter 16

Quantitative Trait Loci

Introduction

Along with aims of sustainable breeding comes their impact assessment. This means that how to quantify, measure and document the sustainability in cattle and buffalo. According there are three approaches : First is a molecular approach to detect and map the quantitative trait loci for performance related traits in cattle and buffaloes. The animals which excel in terms of performance should be first screened in a herd and then selected to increase their progeny at both the organised farms and in the field conditions. Most economically important traits in livestock are affected by many genes as well as the environment and their interactions. If the animal population shows that these economic traits are present among certain animals then there is a need to identify a sustainable livestock with the established markers in the genome. This approach is known as quantitative trait loci approach and is dealt in detail in this chapter. Second approach is statistical approach or multivariate analysis. Third is a documentation of livestock and to evaluate the sustainability. The second and third approaches are dealt in detail from chapter 17 to 19 of this book.

Quantitative Trait Loci (QTL)

QTL is a region on a chromosome, detected via statistical analysis that harbours a gene or genes that influence the phenotypic expression of a complex, quantitative and economically important trait. Hence, a QTL is defined as a polymorphic locus which contains alleles that differentially affect the expression of a continuously distributed phenotypic trait. Usually it is a marker described by statistical association to quantitative variation in the particular phenotypic trait that is thought to be controlled by the cumulative action of alleles at multiple loci. More than one QTL for a particular trait can be present on a chromosome and in the entire genome.

Utility of QTL Mapping

The availability of highly polymorphic DNA markers on genetic maps in different livestock species and their association with phenotypes gave geneticists and breeders an effective tool for the detection of QTL affecting traits of interest. The purpose of mapping QTL in livestock species is to identify genes affecting a quantitative trait and ultimately use existing variation in those genes to select superior individuals from a population. A major objective of QTL studies is to find genes/markers that can be implemented in breeding programs via Marker Assisted Selection (MAS). Once mapped, marker assisted selection for specific QTL can be more effective than mass selection particularly for low heritability traits. There is general agreement from theoretical and simulation studies that application of MAS has the potential to increase the rate of genetic gain especially if traditional selection is less effective (low heritable traits or traits for which measurement of phenotype is difficult, expensive, only possible late in life, or not possible on selection candidates).

In dairy cattle, MAS could be used to pre-select young candidate bulls prior to progeny testing, thus increasing selection differentials, shortening generation interval and increasing genetic gain. Once a QTL is identified, it is necessary to identify families in the breeding population which are segregating for that QTL. However, if a QTL has been fine mapped with respect to closely linked markers that are in linkage disequilibrium (LD) with the QTL, the associations between specific marker haplotypes and QTL alleles should hold across populations and need not be re-established for each individual family. Selection for such QTL can be undertaken throughout the population rather than only in the specific families, thereby greatly simplifying the implementation of MAS. So identification of genes underlying QTL can provide the most accurate markers for MAS.

Marker-assisted introgression of specific mapped QTL can also be an effective means of upgrading native breeds while retaining adaptation to local conditions. QTL mapping will be an essential step in the cloning of QTL by application of the "reverse" genetic procedures.

Resources for Mapping QTL

The basic resources critical to mapping of QTL are appropriate pedigreed populations with production records and genomic DNA samples. Both partial and full genome scans for QTL have been conducted on a number of dairy cattle populations using grand daughter designs (described shortly below). One such population is the Dairy Bull DNA Repository (DBDR) maintained at the Laboratory of Immunogenetics, Department of Animal Sciences, University of Illinois at Urbana-Champaign, which is a collection of Holstein Friesians dairy bull semen and DNA. DBDR has been extensively used for QTL detection using the granddaughter design. Most of the DBDR sires were used in the 1980s and so this population may not be representative of the present population. A new population termed the Cooperative Dairy DNA Repository (CDDR) is being created for analysis of current generations.

Detection of QTL

Due to advances in genomic technology and research, detection of QTL has become the first step in the identification and further study of the gene(s) that actually influence the expression of the trait being studied. In laboratory animals and in plant, the most efficient QTL detection designs involve crosses of inbred lines. The design is equivalent to one large family as the parental lines are completely homozygous. In most domestic animal species, however, no complete inbred line is available and QTL detection should be carried out within-family. Under these conditions, a key factor for QTL detection is the number of progeny per parent. In some species, as in cattle, female prolificacy is low and large dam families are very difficult to obtain. Consequently, only sire families efficiently contribute to QTL detection. In this context, two designs exist for detecting QTL in livestock species. These are daughter design and granddaughter design. In these designs, maternal meioses, i.e. half the total number of meioses, are not used and this strongly affects their detection power. Consequently, these designs are typically 4 to 10 times larger than those involving inbred lines. For a daughter design, genotypic information is recorded for sires and their daughters, with phenotypic observations made on the daughters. For a granddaughter design, the grandsires and sires are genotyped, and phenotypic observations are made on the granddaughters. The power of QTL detection is more in a granddaughter design over daughter design as a result of highly accurate estimates of the breeding values of the sires can be obtained.

Daughter design

Principle

A parent heterozygous for both a marker locus and a linked QTL produces progeny that inherit different alleles for the marker. As the marker and QTL are linked, it is assumed that recombination do not occur between marker and QTL. So offspring receiving a particular marker allele from the parent will also tend to receive the linked QTL allele. Consequently, when progeny are grouped according to the marker allele received from a heterozygous parent, the presence of alternative alleles at the linked QTL will tend to generate a difference in mean value of the quantitative trait concerned between the two progeny groups. Conversely, a significant difference in mean quantitative value of progeny groups receiving alternative marker alleles from their common parent will be indicative of the presence of a linked QTL near the marker. In dairy cattle populations large numbers of progeny can be obtained from a sire, and accordingly sire-daughter families could be used to map QTL.

Model used

A hierarchical ANOVA can be applied in which daughter subgroups receiving the same marker allele from their sire are nested within sires. The analysis model will be:

$$Y_{ijk} = S_i + M_{ij} + e_{ijk}$$

Where,

Y_{ijk} = Trait value for daughter 'k' of sire 'i' that received the marker allele 'j'

S_i = Effect of sire 'i'

M_{ij} = Effect of marker allele 'j' for sire 'i'

e_{ijk} = Random residual error associated with each record.

The analysis tests the significance of the component of variance attributed to marker alleles. A quantitative effect associated with the marker alleles due to QTL linkage will show up as a significant component of variance between daughter subgroups receiving alternative marker alleles from their sires.

Grand daughter design

Principle

In this design a number of sons of a heterozygous sire would be screened for their genotype with respect to the genetic markers and granddaughters are evaluated for the quantitative traits. Therefore this type of analysis is termed as granddaughter design. The heterozygous parent in this design will be the grandsire. The advantage in this design is that it may be easier to collect blood or semen samples from sons of sires (for genotype screening) concentrated in AI centers, than from their daughters, scattered over many farms.

Model used

The model for the granddaughter design will be:

$$Y_{ijkl} = G_i + M_{ij} + SO_{ijk} + e_{ijkl}$$

Where,

Y_{ijkl} = Trait value of the granddaughter 'l' daughter of sire 'k' (son of grandsire 'i') that received the marker allele 'j'

G_i = Effect of grandsire 'i'

M_{ij} = Effect of marker allele 'j' of grand sire 'i'

SO_{ijk} = Effect of son 'k' (of grandsire 'i') with marker genotype 'j'

e_{ijk} = Random residual error associated with each record

In this model sons receiving the same marker allele from the "grandsire" are nested within grandsires. This is a three level hierarchical ANOVA, sons are nested within marker alleles, which in turn are nested within grandsires. As in the daughter design, the analysis tests the significance of the component of variance attributed to marker alleles.

A significant marker-within-sire effect (daughter design) or marker-within-grandsire effect (granddaughter design) is then indicative of a segregating QTL linked to the marker. In both the models residuals are assumed to be distributed independently and identically normal. The daughter design would require between 5 and 20 sires with 200 to 2000 daughters each. This type of analysis might be feasible in situations where the number of proven bulls used in any single year is low, herd size is large, and distances between herds are small. The granddaughter analysis might be more appropriate in situations where elite sires can have many progeny-tested sons.

Candidate gene approach

The objective of breeders is to exploit the genetic variability present among cattle & buffaloes by selecting those with a superior genetic potential for the trait of interest. The most strategy for going from mapped QTL region to gene is the candidate gene approach. A candidate gene for a trait is a gene for which there is some a previous basis in terms of information is available for suspecting that the gene affects the trait i.e. candidate gene known to function in the biochemical pathway that affects the trait. These are two main ways of the candidate gene approach i.e. *comparative and physiological.* Comparative approach takes, as starting point loci where polymorphisms are known to have a phenotypic effect in one species and explores them as candidates for similar variation in other species. The physiological candidate strategy focuses on polymorphism in, or close to genes with functions that relate directly to the trait of interest. An association between a candidate gene polymorphism and the trait of interest is taken as evidence that the gene involved in the genetic control of the trait. Although the polymorphism used as marker may not itself directly control variation in the trait, it is assumed that the functional polymorphism is sufficiently close to the marker that the two loci are tightly linked and in linkage disequilibrium.

Discovery of DNA-microsatellites that are highly polymorphic has stimulated research on detection and mapping of QTL in animal species. Many studies have been undertaken on the use of DNA-microsatellites (genetic markers) in the selection of animals in breeding programmes by a process known as marker assisted selection (MAS). Implementation of MAS essentially involves two steps: identification of marker loci linked to QTL of economic importance and then utilization of linkage association in genetic improvement programme. Once linkage between a QTL and a marker locus is established by testing for statistical associations, it is possible to recognize the alternative QTL allele inherited by the individual. Such information then can be used for the selection of breeding stock. For example, consider a hypothetical situation where a molecular marker M (with two alleles M_1 and M_2), that we can identify using a DNA assay, is known to be located on a chromosome close to a gene of interest say milk protein gene with a variant V_1 that increases the protein yield in milk and V_2 that decreases the yield, that is, as yet, unknown. Then, if a given individual in the population has the allele M_1V_1 on one chromosome and M_2V_2 on the other chromosome. We know that any of its progeny receiving the M_1 allele will have a high probability (how high it is depends on how close M and V are to each other on the chromosome) of also carrying the favourable V_1 allele, and thus would be preferred for selection purposes, while those that inherit M_2 allele will tend to have inherited the unfavourable V_2 allele, and so would not be preferred for selection. With conventional selection, relying on phenotypic values it is not possible to use this kind of information.

MAS is likely to complement rather than replace the conventional breeding systems leading to increased rate of genetic improvement through higher selection intensity, reduction of generation interval and increase in the accuracy of prediction. Although most of the scientists agree that a combination of marker information and phenotypic records is superior to either alone, the weighting of the two sources of data is under investigation. Selection based on markers is possible in early life or in individuals of both sexes for sex limited traits. However, there is a risk of reduced genetic response if the marker association information is inaccurate as MAS is a form of indirect selection. Table (21) outlines few of the identified QTL in European cattle particularly the Holstein Friesian.

The success of MAS is influence by the relationship between the markers and the genes of interest. As suggested by Deckkers (2003), three types of relationship exists

(i) The molecular marker is located within the gene of interest, and such markers are most difficult to find and establish their relationship.

(ii) The marker is in linkage disequilibrium (LD) with allele of interest throughout the whole population. LD is the tendency of certain combinations of alleles (as M_1V_1 in above example) to be inherited together. Population wide LD can be found when markers and genes of interest are physically very close to each other and or when two lines / breeds have been crossed in recent generations. LD can occur by chance in small populations.

(iii) The marker is not in linkage disequilibrium means it is in linkage equilibrium (LE) with gene of interest throughout the whole populations. This is the most difficult situation for applying MAS. The association between the markers and the QTL is a function of distance between the markers and target traits, type of linkage phase and degree of linkage disequilibrium. Therefore, a high-density gene map with closer linkage is a prerequisite for successful implementation of MAS.

QTL Database

The Animal Quantitative Trait Loci database (Animal QTLdb) is designed to house all publicly available QTL data on livestock animal species from which researchers can easily locate and compare QTL within species. Currently, this database contains data on 1287 pig, 630 cattle and 657 chicken QTL (http://www.animalgenome.org/QTLdb/).

The Cattle Quantitative Trait Loci database (CattleQTLdb) has gathered all cattle QTL data published in the past. The release of the CattleQTLdb contains 846 QTLs from 55 publications. Those QTLs represent 91 different traits. These numbers are increasing day by day. On can visit the website, for CattleQTLdb development on following URL:

http://www.animalgenome.org/QTLdb/cattle.html

http://sol.marc.usda.gov

http://spinal.tag.csiro.au

http://locus.jouy.inra.fr

http://www.ri.bbsrc.ac.uk/bovmap

Table 21: QTL Identified for Production & Disease Traits in Cattle

Trait	Chromosome(s) identified
Milk production	1, 6, 9, 10 and 20
Milk yield and composition	1, 2,3, 7 and 14
Mastitis	5, 22, 14 and 23
Trypno-tolerance	23
Muscular hypertrophy	2
Marveling	27
Ovulation rate	5

Softwares Available for QTL Mapping

MultiQTL ver. 2.5
QTL cartographer
MapQTL ® 5
Mapmaker/QTL
MAPMAKER/EXP
Mapmanager
QU-Gene
PLABQTL
QTL express
Genetic power calculator

Summary

QTLs are useful starting points for marker-assisted breeding, positional cloning, functional and evolutionary studies. However, only the QTLs of largest effect and those closest to a marker locus, will show statistically reliable associations. It may be difficult to estimate even the presence of QTL if they interact strongly in their effects and the regions to which a QTL is localized can be quite large (several cM, where 1 cM can range from hundreds to thousands of kilobases). Such regions may contain many genes, and there is no guarantee that a QTL will correspond to only one gene. Thus, with QTLs in hand, much further work is necessary to truly dissect quantitative variation at the mechanistic level.

The candidate approach relies on the presence of linkage disequilibrium within a population at the level of the gene. If there is no linkage disequilibrium, no associations are found (unless the polymorphism detected by the marker itself causes variation). If linkage disequilibrium is wide spread then we may detect associations between a candidate gene and variation controlled by a locus that is some distance away. In conclusion, candidate gene approaches require many independent validation analyses to verify a candidate gene as the QTL affecting traits of interest.

Chapter 17

Multivariate Analysis

Introduction

Animal breeders are frequently planning multivariate analysis using multivariate models which can be substantiated with the growing publications in the field of animal breeding. Generally, for animal improvement programme, we deal with multi traits which are usually inter-correlated to varying degrees and therefore, separate single trait analysis can often misleadingly over estimate or under estimate the time dimensionality of divergence, hence the correct approach must involve the appropriate multivariate analysis to find out the nature of such relationship among the traits and how effectively we can use the results for sustainable breeding and to improve performance of the animals.

One of the important features of using multivariate analysis is whether the characters of the animals represents sample from the single population or many populations. Thus, the generated data matrix may reflect the measurements / records of several characters on individual animals from a single population of a particular farm / locality or on cattle & buffaloes from several farms / localities of the one or different populations.

As the name indicates, multivariate analysis comprises a set of techniques dedicated to the analysis of data sets with more than one variable. Several of these techniques were developed recently in past because they require the computational capabilities of modern computers. Also, because most of them are recent, these techniques are not always unified in their presentation, and the choice of the proper technique for a given problem is often difficult. Moreover, the use of multivariate analysis in animal breeding research is strenuous in nature and needs simulations to derive the interpretation. Thus, it would be very difficult but not impossible as well as time consuming to conclude the analysis of various types of multivariate analysis without the help of specific computer software.

Types of Multivariate Analysis

There are many types of multivariate analysis which can be framed out in animal breeding research programme based on the involvement of one or more than one population as follows

1. Multiple Regression Analysis
2. Principal Component Analysis
3. Discriminant Analysis
4. Genetic Divergence Analysis
5. Canonical Variate Analysis

To cover the various types of multivariate analysis within specified words may not be justified for the beginners involved in multivariate research, however, the authors will to delineate in very brief manner the principles of various multivariate analysis without discussing in detail the statistical procedures of each type of multivariate analysis.

Multiple Regression Analysis

The term regression was coined by F. Galton in 1886 for the study of two variable which known as simple regression. When more than two variables are taken it is known as multiple regression. The regression model may be linear or non linear. The main purpose of multivariate regression analysis is to assess the percent increase or decrease of dependent character by unit increase or decrease of independent character keeping no change of the other independent character(s) of the animal. The technique is specified for a single population and the principle involves the least square analysis.

Concept of multiple regression analysis

The principle of least square analysis is "Sum of Square of deviation between observed and expected variables should be least or sum of square of error should be least". The regression coefficients are estimated through the regression analysis which can be defined as the "amount which quantifies the change in dependent variable and estimated through regression model is called as regression coefficient". The significance of regression coefficient is tested by t-test at error degrees of freedom.

The multiple regression analysis is done with certain assumptions are made:

1. The relationship between the dependent and independent characters is linear in nature.
2. Dependent and Independent variables should be continuously distributed.
3. Error should be normally and independently distributed with the mean zero and unit variance.

$$e \sim N(0, \sigma^2_e)$$

Thus, the general linear multiple regression models will be

$$Y_1 = \beta_0 + \sum_{i=1}^{p} \beta_i X_i + e_i$$

Where, P is more than one character.

For non linear case, the model may be polynomial, multiplicative, exponential or reciprocal (Pindyck and Rubinfeld, 1981). The best equation is to be judged in each method of multiple regressions by using criteria like R^2, mean sum of squares due to error and Mallows Cp Statistic (Mallows, 1973)

Use / Application of multiple regression analysis

1. Regression is used to denote estimation of prediction of the average value of dependent variable for a specified value of independent variables.
2. Used in correction of data.

Principal Component Analysis

The Principal Component Analysis is mainly confined to single population and the main purpose of this study is to reduce the dimensionality of the original data while minimizing loss of information. For the first time, the research on Principal Component Analysis in India was reported in large animals for predicting the breeding value of milk production on the basis of selected number of Principal Components by Dhara and Chakravarty, 1996.

Principal components analysis is a mathematical technique given by T. Hotelling (1933) which describes a multivariate set of data using derived variables. The derived variables are formulated using specific linear combinations of the original variables. The derived variables are uncorrelated and are computed in decreasing order of importance; the first variable accounts for as much as possible of the variation in the original data, the second variable accounts for the second largest portion of the variation in the original data, and so on. Principal components analysis attempts to construct a small set of derived variables which summarize the original data, thereby reducing the dimensionality of the original data. The principal components of a multivariate set of data are computed from the eigen values and eigenvectors of either the sample correlation or sample covariance matrix. If the variables of the multivariate data are measured in widely differing units (large variations in magnitude), it is usually best to use the sample correlation matrix in computing the principal components; another alternative is to standardize the variables of the multivariate data prior to computing principal components. Standardizing the variables essentially makes them all equally important by creating new variables that each have a mean of zero and a variance of one. Proceeding in this way allows the principal components to be computed from the sample covariance matrix.

Concept of principal components analysis

Multivariate data consist of observations on several different variables for a number of individuals or subjects. Data of this type arise in all the branches of science, ranging from psychology to biology, and methods of analyzing multivariate data constitute an increasingly important area of statistics. Indeed, the vast majority of data in animal science is multivariate and proper handling of such data is highly essential.

Most of the times the variables under study are highly correlated and as such they are effectively "saying the same thing". To examine the relationships among a set of *p* correlated variables, it may be useful to transform the original set of variables to a new set of uncorrelated variables called *principal components*. These new variables are linear combinations of original variables and are derived in decreasing order of importance so that, for example, the first principal component accounts for as much as possible of the variation in the original data.

Let $x_1, x_2, x_3, \ldots, x_p$ are variables under study, then

First principal component may be defined as:

$$z_1 = a_{11} x_1 + a_{12} x_2 + \ldots\ldots + a_{1p} x_p$$

Such that variance of z_1 is as large as possible subject to the condition that

$$(a_{11})^2 + (a_{12})^2 + \ldots\ldots + (a_{1p})^2 = 1$$

This constraint is introduced because if this is not done, then Var (z_1) can be increased simply by multiplying any a_{1j}s by a constant factor. The second principal component is defined as

$$z_2 = a_{21} x_1 + a_{22} x_2 + \ldots\ldots\ldots + a_{2p} x_p$$

Such that Var (z_2) is as large as possible next to Var (z_1) subject to the constraint that

$$(a_{21})^2 + (a_{22})^2 + \ldots\ldots + (a_{2p})^2 = 1$$

and cov $(z_1, z_2) = 0$

It is quite likely that first few principal components account for most of the variability in the original data. If so, these few principal components can then replace the initial p variables in subsequent analysis, thus, reducing the effective dimensionality of the problem. An analysis of principal components often reveals relationships that were not previously suspected and thereby allows interpretation that would not ordinarily result. However, Principal Component Analysis is more of a means to an end rather than an end in itself because this frequently serves as intermediate steps in much larger investigations by reducing the dimensionality of the problem and providing easier interpretation. It is a mathematical technique which does not require user to specify the statistical model or assumption about distribution of original variates. It may also be mentioned that principal components are artificial variables and often it is not possible to assign physical meaning to them. Further, since Principal Component Analysis transforms original set of variables to new set of uncorrelated variables, it is worth stressing that if original variables are uncorrelated, and then there is no point in carrying out principal component analysis.

Objectives of principal component analysis

Among the objectives of Principal Components Analysis are the following.

1. Dimensionality reduction
2. The determining of linear combinations of variables
3. Feature selection: the choosing of the most useful variables
4. Visualization of multidimensional data
5. Identification of underlying variables
6. Identification of groups of objects or of outliers

The tasks required of the analyst to carry these out are as follows:

1. In case of a table of dimensions n x m, each of the *n* rows or objects can be regarded as an *m*-dimensional vector. Finding a set of m′ < m principal axis allows the objects to be adequately characterized on a smaller number of artificial variables. This is advantageous as a prelude to further analysis as the m - m′ dimensions may often be ignored as constituting noise; and, secondly, for storage economy (sufficient information from the initial table is now represented in a table with m′ < m columns). Reduction of dimensionality is practicable if the first m′ new axis account for approximately 75% or more of the variance. There is no set threshold, – the analyst must judge. The cumulative percentage of variance explained by the principal axis is consulted in order to make this choice.

2. If the eigen value is zero, the variance of projections on the associated eigenvector is zero. Hence the eigenvector is reduced to a point. If this point is additionally the origin (i.e. the data is centered), then this allows linear combinations between the variables to be found. In fact, we can go a good deal further: by analyzing second-order variables, defined from the given variables, quadratic dependencies can be straight forwardly sought. This means, for example, that in analyzing three variables, y_1, y_2, and y_3, we would also input the variables y_{12}, y_{22}, y_{32}, y_1y_2, y_1y_3, and y_2y_3.

If the linear combinations $y_1 = c_1 y_{22} + c_2 y_1 y_2$ exists, then we would find it. Similarly we could feed in the logarithms or other functions of variables.

3. In characteristic trait selection we want to simplify the task of characterizing each object by a set of attributes. Linear combinations among attributes must be found; highly correlated attributes (i.e. closely located attributes in the new space) allow some attributes to be removed from consideration; and the proximity of attributes to the new axis indicate the more relevant and important attributes.
4. In order to provide a convenient representation of multi-dimensional data, planar plots are necessary. An important consideration is the adequacy of the planar representation: the percentage variance explained by the pair of axis defining the plane must be looked at here.
5. Principal component analysis is often motivated by the search for latent variables. Often it is relatively easy to label the highest or second highest components, but it becomes increasingly difficult as less relevant axis is examined. The objects with the highest loadings or projections on the axis (i.e. those which are placed towards the extremities of the axis) are usually worth examining: the axis may be characterisable as a spectrum running from a small number of objects with high positive loadings to those with high negative loadings.
6. A visual inspection of a planar plot indicates which objects are grouped together, thus indicating that they belong to the same family or result from the same process. Anomalous objects can also be detected, and in some cases it might be of interest to redo the analysis with these excluded because of the perturbation they introduce.

Uses / Applications of principal components analysis

1. The most important use of principal component analysis is reduction of data.
2. It provides the effective dimensionality of the data. If first few components account for most of the variation in the original data, then first few components' scores can be utilized in subsequent analysis in place of original variables.

3. Plotting of data becomes difficult with more than three variables. Through principal component analysis, it is often possible to account for most of the variability in the data by first two components, and it is possible to plot the values of first two components scores for each individual. Thus, principal component analysis enables us to plot the data in two dimensions. Particularly detection of outliers or clustering of individuals will be easier through this technique. Often, use of principal component analysis reveals grouping of variables which would not be found by other means.

4. Reduction in dimensionality can also help in analysis where number of variables is more than the number of observations, for example, in discriminant analysis and regression analysis. In such cases, principal component analysis is helpful by reducing the dimensionality of data.

5. Multiple regressions can be dangerous if independent variables are highly correlated. Principal component analysis is the most practical technique to solve the problem. Regression analysis can be carried out using principal components as regressors in place of original variables. This is known as *principal component regression*.

6. In case of a table of dimensions n x m, each of the n rows or objects can be regarded as an m-dimensional vector. Finding a set of $m' < m$ principal axis allows the objects to be adequately characterized on a smaller number of artificial variables. This is advantageous as a prelude to further analysis as the $m - m'$ dimensions may often be ignored as constituting noise; and, secondly, for storage economy (sufficient information from the initial table is now represented in a table with $m' < m$ columns). Reduction of dimensionality is practicable if the first m′ new axis account for approximately 75% or more of the variance. There is no set threshold, – the analyst must judge. The cumulative percentage of variance explained by the principal axis is consulted in order to make this choice. This is a task for analyst as well as advantage of principal component analysis.

Discriminant Analysis

Concept of discriminant analysis

The study of discriminant analysis is generally done when the number of population is more than one and each animal in the population has multiple characters. The main purpose of discriminant analysis is to find the discriminant function (D) as suggested by Fisher (1936) that increases the differences among populations by minimizing the variance within the population and maximizing the mean differences between the populations with respect of characters. Thus, assuming three characters the linear model for discriminant function we get as

$$D = \lambda_1 d_1 + \lambda_2 d_2 + \lambda_3 d_3$$

$$D = \Sigma \lambda_i d_i$$

Where,

d_i is the ith mean difference of the populations in relation to character, and

λ_i is the weighting coefficient attach to the difference.

Uses / Applications of discriminant analysis

1. Application of discriminant function is that it is used as a basis for making selection on several characters simultaneously. Discriminant analysis aims at discriminating the desirable genotypes from undesirable ones on the basis of their phenotypic performance.
2. The Discriminant Function is utilized in the construction of multi trait selection indices.
3. For classifying the two populations as low milk yielding and with higher milk yielding animal populations.
4. To develop a Discriminant Function to differentiate diseased animals from the healthy animals. Or, differentiate more fertile animals from less fertile / infertile animals.

Genetic Divergence Analysis

Concept of genetic divergence analysis

In any animal breeding programme, the selection of male as well as female animals has utmost interest for successful implementation of the programme in the herd. The purpose may be many but to achieve the goal of distinguishing one animal genetically from the other needs the evaluation of the animal based on many characteristics in different stages to their life. This is perhaps because of genetically divergence of animals with respect to the expression of their characteristics in different levels such as phenotypic, milk protein, gonadal, blood and biochemical, immunological, chromosomal and molecular level. Thus, the use of genetic divergence analysis has an obvious importance for differentiating the similarities and differences of animals. The genetic divergence analysis is basically a multivariate technique and several workers have modified the method from time to time. The significant genetic differences among animals lead to the estimation of genetic distance values (D^2-values) by summing up the squares of deviation of the same transformed and or untransformed traits between the two genotypes in various combinations as:

$$D^2 = \sum_{i=1}^{p} d_i^2$$

$$D^2 = \Sigma (Y_i^j - Y_i^k)^2$$

Where,

i = no. of traits varying form 1 to p

j & k are the genotypes while $j \neq k$

Genetic distance analysis was given by P.C. Mahalanobis (1928). A study on dairy crossbred animals was used to find out whether Karan Fries cows within a particular level of exotic inheritance (upto 50%, 51 - 62.5% and 63 - 75%) were similar or different with respect to level of heterozygosity and 305 days first and second lactation milk production as reported by Sinha *et al.*, 2000.

Uses / Applications of genetic divergence analysis

Genetic distance studies are applied to find the genetically similar or dissimilar animals. For example, preliminary selection of bulls is done via. genetic distance studies before putting them for bull evaluation / progeny testing. The bulls from different genetic clusters must be chosen, if possible, so that a minimum chance of inbreeding occurs when such proven bulls are utilized in breed improvement programmes.

Canonical Variate Analysis

Concept of canonical variate analysis

The Canonical Variate Analysis is mainly used in single population where the traits of the animals can be logically divided into two sets and the relationship between two sets is to be evaluated. The technique was given by Hotelling (1935). The technique obtains the linear combination of one set of characters (criteria character or response character, Y set) that maximally correlated with the linear combination of other set of characters (predictor characters, X-set). Assuming two linear models which are the canonical variates / functions of two sets of characters as

$$M = P_1 Y_1 + P_2 Y_2$$

$$N = q_1X_1 + q_2X_2 + q_3X_3$$

We have to determine the set of canonical weights or canonical coefficients (p_1, p_2 and q_1, q_2 and q_3) in such a way so that the correlation between the two sets of the characters or canonical variates / functions (M & N) becomes maximum and that correlation is called Canonical Correlation. Canonical Variate Analysis has been used for the first time in our country in dairy buffaloes by Thomas and Chakravarty (1999)

Uses / Applications of canonical variate analysis

This technique is applied where more than one correlated variables need to be predicted simultaneously.

□□□

Chapter 18

Documenting Genetic Resources for Assessing Sustainability

The documentation of the genetic resources was initiated around 9000-5000 BC. During 5000 BC-1700 AD, the documentation is also available on migration of human populations with their domestic animals and the number of animal breeds adapted in different environment. Around 1700-1945 AD, the documentation is also available on controlled mating; preparing and keeping of herd books for registrations of animals and choice of breeds for more intensive selection for breeding purpose. Since 1945 the documentation has been made more scientific on various aspects of animal genetic resources for assessing sustainable biodiversity in farm animals. Till now, there are about 3213 number of breeds or breed varieties for domestic livestock which have been documented including 861 cattle (27%), 74 water Buffalo (2%), 427 horse (13%), 1048 sheep (33%), 373 goat (12%), 355 pig (11%) and 75 ass (2%). These include 1219 (38%) for Europe, 856 (27%) for Asia, 423 (13%) for Africa, 322 (10%) for the former USSR, 266 (6%) for North and Central America, 118 (4%) for South America and 69 (2%) for Oceania.

Reasons for Documentation

It is a hard fact that continuous global changes are tending to deplete biological diversity. But why should efforts is made to preserve animal genetic resources? One group argue that there is no need to be concerned since automatic economic and market adjustments will take care of global genetic diversity. On the other hand some feel that the loss of existing unique genetic material is unacceptable. Animal genetic resources are part of the biological diversity of the planet. Their genetic variation is being threatened from several sources.

In developed countries the animal genetic resources are under threat as a result of intensification of agriculture and animal production and greater reliance of those countries on a few breeds and the neglect of the remainder. Some of the countries even identified the National Breeds at the cost of their vast genetic resources. In developing countries the greatest threats to animal genetic resources are the increased and indiscriminate use of artificial insemination, intensification of agriculture and perhaps indiscriminate crossbreeding of local breeds. It is estimated that 35% of mammalian breeds and 63% of avian breeds are at risk of extinction and thereby not sustainable. The performance, adaptation and disease resistance of the vast majority of breeds has not been systematically recorded for assessment of sustainable biodiversity. The majority of livestock genetic diversity is found in the developing world where documentation is most lacking and risk of extinction is highest and increasing. In marked contrast to the situation with plants, wild ancestors of livestock either no longer exist or represent fairly limited gene pools. Lacking the essential information on the genetic resources that are available, most livestock improvement programs have been either non sustaining or unable to strategically target the most appropriate genetic resources, leading to inefficient use of scarce development funds. In this situation the scientific documentation has a key role to play in listing and describing the breeds that currently exist, in documenting the erosion of these resources and in highlighting the breeds that are likely to disappear or not sustainable in the near future.

Sources of Information for Documentation

The information for documentation has to be based on the following sources

1. Dictionary on documented breeds

The dictionary of Masson can be used as the primary information source for all the breeds and breed varieties existing or thought to exist worldwide.

2. Scientific literature

The existing methods of accessing information of animal genetic resources are mainly based on various kinds of publications. Many such publications have been printed over the years and have focused on the animal genetic resources of large countries, regions and species that are of special interest. Such publications soon become expensive to update and reissue. Thus, new information is frequently not accessible. A survey carried out by FAO and UNEP in 12 developing countries showed that only 25 percent of the locally available or nationally published papers, articles, theses and surveys on indigenous animals have ever appeared in the international abstracting services of Animal Breeding Abstracts and AGRIS (International Information System for the Agricultural Sciences and Technology). This is a severe restriction of valuable information and can lead to repetitive work being carried out later.

3. World watch list

The periodic publication of the World Watch List, which will include the latest data sent by national institutions via Regional Animal Gene Banks in to the Global Animal Genetic Data Bank, should ensure better use of information on both animal genetic resources and census data.

Documentation for Maintaining Sustainability

The flow of data from participating countries will permit a regular appraisal of the status of breeds. This can only be done effectively at a global World Watch List Centre since few breeds exist in one country alone. The analysis of data will focus on the population census. Some authors have estimated the minimum number of breedable females required to sustain a breed in different species as given below

Species	No. of breeding females	
	Alderson (1981)	Maijala (1984)
Cattle	750	1000
Horses	1000	-
Sheep	1500	500
Goats	500	200
Pigs	150	200

Recently FAO uses a working rule that when a population size falls to 5000 breeding females and a total population of about 10000 animals, the survival risk of the breed should be studied and appropriate actions are to be taken. Action taken will depend on the local circumstances of the breed, the management system, the extent of cross-breeding, the rate of decline in numbers and the certainty that the breed has unique qualities. According to established principles, specific recommendations will then be made for each circumstance.

Type of Documentation Required for Assessing Sustainability

The Global Animal Genetic Data Bank established at the Institute for Animal Breeding and Genetics, University of Hanover, Germany and the Animal Descriptors, developed by FAO and the United Nations Environment Programme (UNEP) in 1986, is flexible and open-ended so that data can be updated and new traits included. The bank is also expected to be of value not only for preservation activities but also for planners, project designers, governments, extension workers, teaching institutions and commercial interests in the improved use of animal genetic resources. A hardcopy output, based on a standard numerical format but which prints in different languages, is being studied so that language will not be a barrier either to input or access. The first inputs from developing countries will flow via the regional animal gene banks, stocked from national sources, and will include the genetic characterizations and population census data. In addition, in 1989 work started on the transfer of such data from the national animal genetic data banks in China and India. It is the objective that other developing countries, having their own national animal genetic data banks, will use the same standards and thus facilitate a uniform global system. The information has been received from a total of 52 countries. The information which can be stored for any breed for assessing sustainability has been summarized with the following major headings-

1. General Information
2. Origin and Development of the Breed
3. Description of Breed
4. Uses and Qualities of the Breed
5. Management Conditions
6. Production Performance
7. Additional Information

Analysis of Documented Information

In 1993, FAO documented about 2047 breeds of which 86 (4.20 %) were found wild, 502 (24.52%) were described as a variety or strain of a breed and 164 (8.01%) were found in more than one country. Of the above total breeds, 35% had some information on population size, 39% had some information on production performance and 30% breeds contained some information on both population size and production performance (Table 22). In 2004, FAO continued to up-date the information of 2928 breeds belongs to eight domestic animal species.

Table 22: Number of Breeds or Breed Varieties in FAO Global Data Bank

Information	Cattle	Buffalo	Horse	Sheep	Goat	Pig	Ass	Total
Number	564	68	247	600	259	245	64	2047
Information on Population Size	257 (46)	29 (43)	88 (36)	236 (39)	81 (31)	79 (32)	7 (11)	777 (38)
Information on Production Performance	232 (41)	34 (50)	57 (23)	267 (45)	119 (46)	80 (33)	3 (5)	792 (39)
Information on Population Size and Production Performance	211 (37)	23 (34)	40 (16)	209 (35)	67 (26)	64 (26)	2 (3)	616 (30)

Figures in parenthesis represent percent of all entries in the data bank

According to analysis in 1993, a total population size of 5000 breedable animals or less was available on 777 breeds only. However, the continuous effort helped to document about 1671 breeds for which population information is available in 2004. Of the 777 breeds in 1993, a total of 221 considered at risk as they were described as rare, endangered or almost extinct or where actual figures were available, they had a total population size of 10, 000 numbers or less. It was further observed that out of 221 breeds at risk at about 130 (58.82%) are available in developed countries and 91 (41.18%) are available in developing countries (Table 23). Out of total breeds threaten to be as endangered about 68 (30.76%) cattle breeds and only one (0.004%) buffalo breed were in danger list. It was also observed that out of 68 cattle breeds at risk at about 32 (47.06%) belongs to only North, Central and South America alone and of the cattle breeds in danger list, about 25 (36.76%) are in developed countries and 43 (63.24%) are developing countries. It has found in 2004 that the breeds at risk have swelled to 448 or almost 28 percent have been classified as being at risk of extinction based on the classification categories. It is assumed that 10-12 percent of these breeds will become extinct over the next year which shows about one breed we may loose in every week.

Table 23: Number of Breeds or Breed Varieties Considered At Risk

Information	Cattle	Buffalo	Horse	Sheep	Goat	Pig	Ass	Total
Developed World	25	0	53	27	12	11	0	130
Developing World	43	1	10	5	6	21	5	91
Total World	68 (29)	1 (88)	63 (236)	32 (81)	18 (79)	32 (7)	5 (777)	221 (257)

Figures in parenthesis represent the entries in the data bank based on information on population size.

Strategically Warning System

A risk assessment method needs to be established covering each breed of different species in every country. Various estimates have been made for minimum number of live animals required to maintain a breed. The FAO Early Warning System will inform governments of which breeds are approaching the danger level by assessing a population's rate of decline over a period of years using different surveys. The Early Warning System will also predict which endangered breeds could be saved through appropriate anticipatory action. Unfortunately, many local breeds are disappearing without it even being known whether or not they have unique genetic adaptive traits. In the case of domestic livestock, there is often no objective documentation on their performance and adaptive abilities. It is for this reason that the Early Warning System is important as a form of global infrastructure to aid rational decision-making. The Ministry of Agriculture , Govt. of India has also continued to identify one of the major thrust area during the 10th Five Year Plan (2002–2007) is to develop the reliable database and management information system for Monitoring of National Project on Cattle and Buffalo Breeding initiated from 2000 in the country where it was emphasized that provision are to be made for quality breeding inputs in breeding tracts of important indigenous breeds so as to sustain breed from deterioration and extinction.

□□□

Chapter 19

An Impact Study on Breeding and Management Interventions Enhancing Dairy Productivity and Its Sustainability

Introduction

Among rural households in India, dairying in general has been carried out as mixed farming on a small scale and in subsistence manner. The milk and milk products thus produced are primarily meant for domestic household consumption. It is, now encouraging to note that there is a perceptible change in the thinking of various categories of farmers towards rearing cattle and buffaloes for commercial dairy production. The small, marginal and landless agricultural farmers are now appreciating the advantages of adopting dairying a vocation as a dependable source of supplementary income as well as self employment generation. In that context, we are very fortunate to have in our kitty quite a sizable number of indigenous and crossbred cattle and the world's best germplasm of buffaloes. The traditional package of practices for breeding, feeding, housing, management and health care usually followed by farmers for rearing dairy animals particularly of high production potentials do not result in further enhancing and sustaining the productivity of animals. There are several gaps/constraints, which may affect the productivity of indigenous and crossbred cattle and buffaloes and make them highly unsustainable. Therefore, there is an urgent need to identify the critical gaps impeding the productivity of animals under rural household condition and to mitigate such gaps through intervening improved technologies / package of practices and bringing awareness among farmers by field demonstration of newly developed technologies/ package of practices for adoption so that the productivity can be increased and sustained. Keeping in view of the various factors responsible for improving and sustaining the productivity of dairy animals, the project under NATP. In NDRI Karnal, Haryana was started with the following objectives.

1. To assess the level of productivity and identify factors/ gaps affecting the productivity of indigenous and crossbred cattle and buffaloes in rural households.
2. To bridge the gaps for sustaining high production under rural household and enhance the total productive life/herd life of animals.
3. To augment the health status of high producing animals and popularization of appropriate package of practices for sustainable high production.
4. To optimize the cost of milk production for sustainability and profitability.
5. To quantify the impact of the improved technologies/package of practices in relation to the level of productivity in rural households.

Major Gaps/ Constraints Identified Affecting the Productivity of Farmer Herds

While conducting Benchmark survey in the project operation area, under four co-operative centres assessing the socio economic status of farmers, resource inventories and the level of productivity of indigenous, crossbred cattle and buffaloes following gaps/constraints affecting the animal productivity were identified.

1. Inadequate availability of superior animal germplasm (frozen semen of high genetic merit bulls). Breeding of village cattle and buffalo herds through natural service by scrub bulls.
2. Lack of breeding and animal health care infrastructure networking i.e. AI facilities, trained inseminators, regular supply of quality frozen semen liquid nitrogen etc.

3. Lack of awareness among farmers for timely detection of heat and insemination of their animals.
4. Absence of animal performance recording.
5. Poor quality roughage and crop residue based animal feeding.
6. Irregular and unbalanced concentrate feeding and lack of mineral supplementations.
7. Lack of knowledge & adoption of scientific herd management practices.
8. Inadequate marketing facilities of animals, milk and milk products.

To mitigate some of these gaps, important technical interventions like breeding, feeding, animal health & management providing basic technical inputs i.e. establishment of A.I and Animal health centre well equipped with technical manpower, ensuring regular supply of superior germplasm of high performance pedigree bulls, providing better breeding facilities and animal health coverage at the door steps of farmers and bringing awareness among farmers through training and field demonstrations were introduced.

Breeding Management Interventions

Through establishment of Artificial Insemination and Animal Health centres providing breeding and animal health cover facilities at the door step of farmers in the adopted villages has increased adoption level of artificial insemination for breeding cattle and buffaloes tremendously through regular supply of superior germplasm in the form of frozen semen of high genetic merit cattle and buffalo bulls. The adoption of breeding animals through artificial insemination with the semen of high genetic merit cattle and buffalo bulls increased up to 100 % in cattle and 60 % in buffalo herds maintained by relatively progressive rural households under four co-operating centres. The improved breeding management interventions resulted in the birth of 5769 cattle and 3335 buffalo calves with higher expected producing ability, which is a great economic asset of rural households. These improved interventions also brought awareness among farmers for proper detection of heat of animals and timely inseminating their animals thus resulting increased and consistent conception rate almost comparable to institutional organized farms, decline in service period and increase in milk productivity. The breeding interventions have further resulted towards identification of high producing elite animals for future breeding as well as developed basic infrastructure network and available trained man power for breed improvement and development programmes under village conditions. Large number of elite crossbred cows up to a record peak milk yield of 54 Kg. in a day and elite buffaloes up to a record peak yield of 17 Kg in a day has been identified as superior genetic resources for propagation of quality cattle and buffalo germplasm.

Feeding Interventions

For improving the quality of available feed resources and feeding rations, particularly high yielding animals. The following feeding interventions have been identified and introduced in 20-25 rural households rearing high yielding animals.

i) Feeding urea treated wheat straw.

ii) Feeding of mineral mixture @ 50-60 gm. per animal per day.

iii) Feeding of high energy ration supplemented with mustard oil @ 100 gm per animal per day.

iv) Feeding of balance ration.

v) Feeding of formaldehyde treated mustard cake as protected protein feeding.

vi) Preparation and feeding silage.

Supplementing mineral mixture and other feeding interventions to lactating animals revealed increasing and sustaining the daily milk yield, improving fertility and general health of cattle and buffaloes. The rural households have adopted the feeding interventions particularly feeding mineral mixture, vegetable oil and balanced feeding for their high producing lactating cattle and buffaloes. As a result of improved feeding practices, anestrous and repeat breeding problems in cattle and buffaloes were considerably reduced after mineral supplementation in feeding. For adopting improved feeding interventions i.e. formulating their own balanced animal feed for feeding, additional concentrate feeding to advance pregnant animals, mineral mixture feeding, silage making and feeding during lean periods and feeding of urea treated wheat straw etc., there has been increase in number of farmers for adoption of above interventions on their own. The adoption level of formulating own concentrate feed among farmers increased from 34.7% to 49%, feeding of additional concentrate to pregnant animals increased from 69.3% to 87.5% and feeding of mineral mixture was adopted by almost all experimental households. The adoption level was relatively higher among rural households who were resource rich, rear high yielding animals on commercial scale under intensive animal production system.

Animal Health Management Interventions

Animal health management package of practices with prophylactic vaccination and deworming reduced substantially the morbidity and mortality of young calves. The other animal health interventions also reduced the incidence of mastitis, repeat breeding and anoestrus cases in cattle and buffaloes in the rural households at different centre. The interventions like udder washing, teat dipping and dry cow therapy were adopted by rural households at different centre, not only improved the udder health of high yielding dairy animals but also brought awareness among farmers for clean milk production under village conditions.

Impact of Technical Interventions

Therefore as a result of introduction of various improved technical interventions, and scientific package of practices and also their practical demonstrations on the premises of rural households following deliverable impact have been achieved.

1. Brought awareness amongst farmers to breed their herds with superior germplasm in the form of frozen semen of high genetic merit cattle and buffalo bulls. This has further augmented the rural households to have high milk yielding and sustainable cattle and buffalo herds.

2. Awareness and adoption of improved feeding and animal health management interventions by farmers will further enhance productive life/ herd life of high producing animals and their economic sustainability.

3. Trained technical manpower availability in the form of village level laid inseminators which will further continue to provide technical inputs to the farmers.

4. Development of progressive farmers/breeder herds as field level research laboratories and demonstration units for benefits of farmers in the surrounding villages.

Future Work

The above mentioned project has created network of cattle & buffalo breeding infrastructure facilities and developed skilled manpower amongst rural farmers and also developed progressive farmers/breeder herds as field level nucleus breeding herds, research laboratories and demonstration units for future propagation and dissemination of superior germplasm, technologies and herd management package of practices under field conditions. Keeping in view of the establishing facilities for artificial insemination, animal health coverage and animal performance recording under field conditions, the genetic improvement programmes in cattle and buffaloes i.e. identification and selection of superior dairy animals of high genetic merit through progeny testing under farmers herds and multiplication of superior germplasm could be planned for future. Development of farmer's co-operatives linked with activities on improving cattle and buffalo production, processing and marketing of milk and milk products could be possible research and development programmes to be undertaken by R&D organizations under different public and private sectors.

□□□

Glossary

AFC

Abbreviated as Age at First Calving, it is one of the most important performance trait which is recorded for analysis purpose and is defined as the age of the female animal at which she gives birth to her first offspring, and is calculated as period from date of birth to date of first calving.

AICRP

All India Coordinated Research Projects. The Indian Council of Agriculture Research had launched a coordinated research project with the objective of evolving high yielding breeds of different species for different utility. For example AICRP on cattle and buffalo was launched in the year 1970 and 1971 for higher milk production.

Allele

An alternative form of a gene. They occupy same locus on homologous chromosome.

Amino acid

Any of a class of 20 molecules that are combined to form proteins in living things is called as amino acid. The sequence of amino acids in a protein and hence protein function are determined by the genetic code.

Ammoniated

Combined or impregnated with ammonia or ammonium compounds

Animal Cloning

When complete animals are obtained from somatic cells of an animal, it is called animal cloning.

Animal Model

When we demonstrate various biological phenomenon pertaining to interrelationships (i.e. cause and effect relation) we may depict them in the form of mathematical expression, such models are called as animal models. They are different from the mathematical models in terms of error, it means unlike mathematical models animal models have errors associated with it. Animal model may be linear or non linear.

Animal Production Systems (APS)

According to FAO, all input-output relationships viz. biological, climatic, economic, social, cultural and political factors, which combine to determine the production of a particular livestock enterprise at a particular location is referred as Animal Production System.

Antioxidant

A compound added to feeds to prevent the oxidative rancidity of fats or a compound which is added to mixture to prevent the oxidation.

Artificial Chromosome

A construction composed of the basic elements of a normal chromosome into which one or more transgene sequences may be introduced. After introduction into the nucleus, this artificial chromosome is replicated and distributed to daughter cells just like a normal chromosome. The transgene expression is more predictable as it is not influenced by integration into an unknown location.

Artificial Insemination (AI)

A reproductive technique through which the frozen semen is deposited artificially in the female reproductive tract, with the help of trained personnel and AI gun. It has many practical advantages over natural service. This technique has made significant contribution in increasing the reproductive efficiency, global transport of germplasm and its multiplication. Better genetic progress can be made through the use of this technique.

Artificial Vagina (AV)

AV is used to collect the semen from bulls at semen stations or bull farms. It comes in different sizes according to its use for different species.

Associated Herd Progeny Testing

Single herd progeny testing could not yielded desired results mainly for small herd size and hence few sires could be tested. To overcome the problem of smaller herd size, it was thought to associate different herds and testing the sires simultaneously based on data on all herds. This lead to increase the no. of sires tested with higher accuracy. However, this posses the problem of GxE interaction which needs to be taken care of during the design of associated herd progeny testing programme.

Autosomes

Autosomes are all the chromosomes except the sex chromosomes; a diploid cell of cattle has 58 numbers of autosomes.

Available Protein

The portion of the crude protein that can be digested by the animal; it can represent the proportion of total protein after deducting the ADF-N fraction of a feed.

Average Daily Gain (ADG)

The average daily live weight increase of a growing animal, usually expressed in kilograms, grams/day.

Balanced Ration

A 24-hour feed allowance that provides an animal with appropriate amounts and proportions of all nutrients required for a given level of performance.

Base Sequence

The order of nucleotide bases in a DNA molecule.

Beef

Beef is the meat produced after slaughter of cattle's.

Block

One or more feeds compressed into a solid form, usually weighing 15-25 kg.

BLUP

Best Linear Unbiased Prediction is a method sire evaluation which provides directly comparable estimates of the average breeding value of groups of animals born in different years. It takes into account the complications of non random mating, sires from more than one herd, environmental trends over time, herd differences for breeding value of dams and bias due to selection.

Bran

Coarse outer grain coating, separated during processing.

Breed

Is the group of individuals having the same phenotypic and physiological characters and can reproduce a fertile offspring on crossing. Breeds are separated by geographical barriers and are found in different locations.

Breeding Efficiency

Breeding efficiency is estimated for the lifetime of a cow or buffalo based on the number of calvings in lifetime

Breeding Objectives

Under each genetic improvement programme some breeding objectives are set to achieve the successful implementation of that programme. The breeding objectives may be basic, strategic or applied in nature.

Breeding Value (BV)

The average value of an individual judged by the mean value of its progeny is called B.V. It is twice that of the transmitting ability and for practical purpose it is taken as the additive genetic variance $V_{A.}$

Buffalo

The domestic buffalo was given the specific name, bubalis by Linnaeus. According to Bohlkon's terminology, it should be called Bubalus arni forma bubalis. The domestic buffalo in different countries is known by different names such as "bhains" in India, "aljamoos" in Arab countries, karabue" or "kwai" in Thailand, "carabao" in Phillipines and "karbo in Malaysia.

By-pass Protein also known as Rumen undegradable protein (RUP)

The portion of ingested protein that resists rumen degradation and is digested directly in the other stomachs or small intestine; generally a more expensive source of protein

By-product

Feed ingredients produced as a result of industrial manufacturing, plant or animal processing (e.g. distillers' grains, beet pulp)

Cake

Vegetable oils are extracted from various oilseeds (soya, rapeseed, sunflower etc) when the oil has been extracted, it leaves behind a product, which is mainly rich in proteins called cake. Cake is used in animal feeding.

Calving Interval

The period between two successive calving and is a measure of reproductive efficiency of dairy animals. For the reproductive traits it is equal to the sum of service period and gestation period. Calving interval is also defined in terms of production traits, which is the sum of lactation length and dry period.

Candidate Gene

A gene that is thought to be responsible (among many genes possible) causes for a diseases or phenotypes. Candidate gene lies in a region that has been identified through genetic mapping. The protein product of a candidate gene may implicate the candidate gene as being the actual disease gene being sought.

Cattle

The animals comprises of a subfamily, known as bovinae, of the family bovidae of horned animals, having 9 subspecies of cattle under genus Bos.

Census (livestock)

Is the procedure to count the heads or number of all the animals of a particular species. It is carried out at every five (5) years interval in livestock.

Chromosomes

The self- replicating genetic structures of cells containing the cellular DNA that bears in its nucleotide sequence the linear array of genes. In prokaryotes, chromosomal DNA is circular, and the entire genome is carried on one chromosome. Eukaryotic genomes consist of a number of chromosomes whose DNA is associated with different kinds of proteins.

Clone

Is a genetically identical animal that can be produced either by embryo splitting (as occurs in nature) / transfer of embryonic blastomeres or somatic cells as donor nuclei.

Cloning

The process of asexually producing a group of cells (clones), all genetically identical, from a single ancestor. In recombinant DNA technology, the use of DNA manipulation procedures to produce multiple copies of a single gene or segment of DNA is referred to as cloning DNA

Complete Feed

A thoroughly blended mixture of different feed ingredients formulated to meet the specific nutrient requirements that allows for greater efficiency in feeding and provides better control of nutrient intake. A complete feed may or may not include the roughage portion of the ration.

Compound Feed

Is the term used for any feed produced by an animal feed manufacturer. It may be a complete or supplementary feed depending upon whether it constitutes the entire food intake or whether it is used as a complement to diet based mainly on fodder, grass, concentrate etc. produced for farm animals.

Concentrate

A classification of feedstuffs high in energy and low in fibre, usually further divided into energy and protein concentrates. Often used interchangeably with supplement (e.g., corn, barley, soybeans).

Congenital

It means deformities or diseases which are present at birth or immediately after the birth in an individual.

Conservation

It means the preservation along with improvement of the genetic potential and management of a breed for use in future. Thus, the conservation includes continued maintenance of the genetic variability.

Conservation Strategies

There are two means of conservation i.e. *in situ* and *ex situ*. Conserving in the form of live animals that exist in nature is *in situ* conservation. In comparison, *ex situ* conservation is achieved by storage of sperms, ocytes, embryos, DNA, Somatic cells and embryonic stem cells.

Contigs

Groups of clones representing overlapping regions of a genome.

Critical Species

Species or breeds are considered to be critical if their population goes less than 100 animals.

Crossbreeding

It is a system of mating where the two established breeds are crossed and the animals thus produced are called as crossbred animals.

Crossbred Animals

Those animals which are produced by crossing two established breeds in order to utilise the breed complimentarily (e.g. high milk yield and disease resistance both in one breed) are known as crossbred animals. Cross breeding with the exotic breeds of cattle with the local or nondescript cows of India is recommended as cross breeding policy for cows.

Crossing Over

The breaking during meiosis of one maternal and one paternal chromosome, the exchange of corresponding sections of DNA, and the rejoining of the chromosomes. This process can result in an exchange of alleles between chromosomes. Compare recombination

Crude Fibre (CF)

Chemical analysis that involves extraction of a ground feed sample with diethyl ether followed by sequential boiling in dilute acid and dilute base. Residue is then burned in a furnace; CF is calculated as the difference in weight of the sample before and after burning. CF has been replaced by ADF and NDF fibre fractions in ruminant feeds but is still reported for monogastric feeds

Crude Protein

Is an estimate of the total protein content of a feed determined by analyzing the nitrogen content of the feed and multiplying the result by 6.25. Crude protein includes true protein and other nitrogen-containing substances such as ammonia, amino acids, nitrates etc.

Cryopreservation

A method of preserving the biological material (sperm, ova or embryo) under extreme low temperature (-196 °c) for latter use is known as cryopreservation. The freezing is carried out in different steps and slowly. Different types of cryoprotective agent are used to prevent the injuries due to formation of ice crystals in the biological material to be stored. The most common cryoprotectant used is Glycerol.

Digestibility

A measure of the apparent extent that a feed or nutrient is digested, usually expressed as a percentage of the amount consumed.

Digestible Energy (DE)

The apparent energy that is available to the animal by digestion, measured as the difference between gross energy content of a feed and the energy contained in the animal's feces (gross energy minus fecal energy).

Digestion

The process of mechanical, chemical and enzymatic breakdown of consumed feeds into smaller components for absorption in the intestine of the animal

Diploid

A full set of genetic material, consisting of paired chromosomes one chromosome from each parental set. Most animal cells except the gametes have a diploid set of chromosomes. The diploid human genome has 46 chromosomes. Compare haploid

Disease Resistance

Is a particularly important attribute of livestock in sustainable low-input livestock production systems. Resistance to infectious diseases is often the critical determinant of such system and improving resistance is considered as a main target for genetic improvement programmes.

DNA

Deoxyribonucleic acid; the genetic material

Dry Matter

Feed residue left after all moisture has been removed by drying (e.g. 100% dry matter).

Dry Matter Intake

Amount of moisture-free feed consumed

Dry Period

The period between dates of dry to the date of next calving

EBV

Estimated Breeding Value. Used to rank animals for a particular trait

Embryo

A fertilised egg in early stages of development / foetal growth i.e. from conception to the eighth week of gestation.

Endangered Animals

Endangered animals those species or breeds are considered to be endangered if their population goes less than 1000 animals.

EPA

Abbreviated as Expected Producing Ability or MPPA given by Lush, 1945. EPA is of the herd is the most important parameter for the ranking, selection and culling of the adult cows and she buffaloes. EPA is used in selection of females.

EPD

Abbreviated as Expected Predicted Difference and is used as parameter in selection of young bulls based on the performance of his ancestors / pedigree. The animals with higher EPD are tested in progeny testing programme under test mating i.e. the young bulls with high EPD are selected for test mating in progeny testing.

Epistasis

The genetic phenomenon of inter allelic interaction in which expression of one gene obscure the phenotypic expression of another gene.

ETT

Embryo transfer is a recent reproductive technique by means of which elite females could be multiplied faster. In this technique 6-8 day old embryo is flushed from the uterus of a donor female and is then transferred into a recipient female or frozen to be transferred at a later date. Another is the lack of commercially available embryos for sale. In cattle, ET is used extensively and creates significant economic and genetic gain for the industry. The acceptance of ET in cattle came about after embryos could be frozen, and more importantly when the evolving technology became simpler, more reliable and inexpensive, allowing technicians and farmers to transfer these embryos successfully into recipient cows. One factor limiting the wide-spread use of ET in Indian cattle and buffalo is cost.

Exotic Breed

Means not originated in our country and has been imported animals / germplasm from other country (western world).

F_1

F_1 is the first generation produced by crossing two parental homozygous lines.

Feed Block : Refer Block

Feed Efficiency

A ratio describing the amount of feed consumed per unit of production (e.g. body weight gain, milk, eggs).

Flushing

In-Vitro method of collecting embryos from the uterus in female animals. It is usually carried out seven days post-heat in cows (six days post-fertilisation).

Forage

Plants or plant parts fed to, or grazed by, domestic animals. Forage may be fresh, dry or ensiled (e.g. pasture, green chop, hay, haylage).

Frozen Semen

Storage of semen in the frozen state implied a complete arrest in the developmental and metabolic process within the sperm cell. The cell need to be held in a state of suspended animation until required for fertilization.

Gamete

Mature male or female reproductive cell (sperm or ovum) with a haploid set of chromosomes (30 for cattle)

GDP

GDP is abbreviated as Gross Domestic Product

GXE Interaction

Means that the genotype and environmental variations do not combine their effects linearly or additively i.e. a phenotype is not simply the sum of the genotype and the environment but another component of variation (GXE interaction) is also associated with it.

Gene Frequency

Gene frequency may be defined as the proportion of loci which are occupied by a particular allele at a given locus in a population.

Gene

It is the fundamental physical and functional unit of heredity. A gene is an ordered sequence of nucleotides located in a particular position on a particular chromosome that encodes a specific functional product (i.e. a protein or RNA molecule)

Gene Mapping

Determination of the relative positions of genes on a DNA molecule (chromosome or plasmid) and of the distance, in linkage units or physical units, between them

Gene Pool

The sum total of all the alleles present in the breedable population.

Gene Transfer

It is an experimental process for modifying the heritable genotypic and phenotypic content of cell by causing the cell to take up and express the gene sequence of purified DNA from donor cell.

Generation Interval

It is the time period between two successive generations with respect to the same stage of life cycle.

Genetics

It is the branch in biology which involves the study of hereditary and variation of the trait under consideration, in a population.

Genome

A complement set of chromosome (haploid) of a species. The genome may be looked at in two different ways. To a cytogeneticist, it represents the haploid set of chromosomes of a diploid organism. Therefore, a polyploid organism will have more than two genomes, which may be identical (autopolyploids) or distinct (allopolyploids). But to a molecular biological genome consists of the total genetic information present in the organism.

Genotype Frequency

It may be defined as how abundant or rare a particular genotype is present in a population at a particular time.

Genotype

The genetic constituent of an individual organism is called its genotype.

Germplasm

Germplasm in animal breeding refers to the semen of male and ova of female animals but is more commonly used for the frozen semen straws.

Gestation Period

Gestation period is the period from conception to the calving.

Grading Up

Up grading is the system of mating where the crossing is done between male from indigenous well defined breeds of cattle and buffalo to the non descript / local females

Gram (g)

1/1,000 of a kilogram (kg) or 1/28 ounce (28 grams = 1 ounce). It is a small unit of weight.

Gross Energy (GE)

Laboratory analysis that measures the total combustible energy in a feed. It is determined by the amount of heat produced when a feed sample is completely burnt in a bomb calorimeter.

Haplotype

A set of closely related genes on a chromosome (or genetic markers) that tend to be inherited together or single unit is called haplotype. It is the genotype for a specific region of a chromosome.

Herd

Herd is a group of cattle or buffalo.

Heritability

In narrow sense is defined as the ration of additive genetic variance to the total phenotypic variance. It is the most important parameter in formulation of any breeding plan.

Heterozygosity

The presence of different alleles at one or more loci on homologous chromosomes.

Hetrosis

It is the phenomenon where the progeny performance increases than the average performance of their parents or better than the best parent performance. It results due to crossbreeding in F1 generation. In subsequent generations it reduces to half than that of previous generation.

Holstein Friesian

An exotic dairy breed which originated in two provinces of Netherlands (North Holland and West Fries land). Large sized body, with black and white colour. Adult body weighs 800 – 850 kg of males and 500 – 550 kg of females. This is the best dairy breed of the world with world record of about 19000 kg in lactation and an average production of 7000 kg with low fat percent of about 3 – 3.5 %. Age at first calving is about 30 months and calving interval of 13-14 months.

Homologous Recombination

A transgene sequence is constructed which matches the sequence of the endogenous target gene. This biases insertion to the target gene. The design of the gene construct determines whether the target gene will be rendered inactive (knocked out) or simply modified.

In vitro

Outside body of living organisms, in a test tube.

In vivo

Within the living organism

Inbreeding

Is the system of mating where animals which are more closely related are mated. Like sire-daughter mating, dam-son mating and full sibs mating. It has deleterious effects on fertility and viability and fitness traits in the progenies. It is opposite of out-breeding.

Indigenous Germplasm

Means the gene pool from the animals / breeds which are native to the country (India).

Inter-se Mating

Is a system of mating where the crossbred animals are crossed in order to maintain the level of exotic inheritance it gives rise to different genetic groups.

Introgression

If a particularly unique characteristic only exists in an exotic population, it usually difficult to breed that trait into other population. Introgressing a gene into a population is done via conventional breeding, which requires an identification of close and tightly bracketed molecular markers / QTL, or transgenic approach is used for its introduction into a population.

In-vitro Fertilization (IVF)

When the union between egg cell and sperm occurs outside the body in a culture vessel, it is known as in vitro fertilization. This involves collection of healthy ova and sperms from healthy females and males, respectively, and their fusion under appropriate conditions *in vitro.* The resulting zygotes may be cultured *in vitro* for a period of time to obtain young embryos, which ultimately are implanted in the uterus of healthy surrogate females to complete their development.

Jersey

Is an exotic dairy breed originated in island of Jersey. Smaller size cattle with reddish grey to brown colour. AFC and CI are almost same to Holstein Friesian and average production is 4500 kg in lactation with average 4.5% of fat.

Karan Fries

Synthetic cattle breed evolved by crossing Friesian x Tharparkar at the National Dairy Research Institute, Karnal. The breeding policy for improvement of the crossbred population is based on inter-se mating to maintain 50 to 75 percent level of exotic germplasm.

Karyotype

A photomicrograph of an individuals chromosomes arranged in a standard format showing the number, size, and shape of each chromosome type; used in low- resolution physical mapping to correlate gross chromosomal abnormalities with the characteristics of specific diseases.

Kb

Kb (kilobase) is a unit of genome and indicates the 1000 base pairs of DNA

Lactation Length

The period between date of calving to date of dry. As a standard for comparisons it is taken as 305 days in cattle's which includes the first 5 days of colostrums secretion.

Lactation Yield

The amount of milk produced by a cow / buffalo in one complete lactation.

Lifetime Production

The sum total of the milk produced by an animal in her all lactations. For practical purpose it is the production of an animal upto to her fourth lactation, considering that about 80% of the total milk comes till 3rd to 4th lactation.

Lignin

A complex polymer bound to cellulose that strengthens plant cell walls but is indigestible to animals.

Linkage Disequilibrium

A state resulting when alleles at two defined loci are linked more frequently than would be expected based on the known allele frequencies and recombination rate between the two loci. Linkage discquilibrium indicates that the two alleles being examined are close physically in genome and are inheriting together. It implies that a group of markers has been inherited co-ordinately.

Linkage Map

A map of the relative positions of genetic loci on a chromosome, determined on the basis of how often the loci are inherited together. Distance is measured in centimorgans (cM)

Linkage

The proximity of two or more markers (e.g., genes, RFLP markers) on a chromosome; the closer together the markers are, the lower the probability that they will be separated during DNA repair or replication processes (binary fission in prokaryotes, mitosis or meiosis in eukaryotes), and hence the greater the probability that they will be inherited together.

Livestock

Livestock means the farm animals raised for use or to sell

Locus

A fixed position in a chromosome, occupied by a gene.

LOD Score

Short for 'Log Odds', a statistical estimate of the gentic linkage between two loci on the same chromosome. It is defined as the $\log_{10}$ ratio of probability that the data would have arisen if the loci are linked to the probability that the data could have arisen from unlinked loci. The conventional threshold for declaring linkage is a LOD score of 3.0

Longevity

The period between dates / of birth to date of death / disposal of an animal.

Mapping, Genome

A genome map may be defined as a detailed schematic description of the structural and functional organisation of all the chromosomes in the genome of an organism. At present, we have mainly three types of maps: genetic or linkage map, cytogenetic map and physical maps.

Marker

An identifiable physical location on a chromosome (e.g., restriction enzyme cutting site, gene) whose inheritance can be monitored. Markers can be expressed regions of DNA (genes) or some segment of DNA with no known coding function but whose pattern of inheritance can be determined

Marker Assisted Selection (MAS)

An indirect method of selection utilized for the traits with low heritability and disease resistance traits. Usually, with the help of molecular markers, this segregates along with the gene of interest in a population.

Mating

The process of crossing male with female animals in a systematic and scientific manner.

Metabolizable Energy (ME)

A measure of the useful energy in a feed, representing that portion of the feed gross energy not lost in the feces, urine and eructated gases.

Methionine

Is sulphur containing essential amino acid, which is required for protein synthesis. It is a limiting amino acid in majority of the ruminant diets. Protected-methionine means resistant to rumen digestion.

Milch Animals

The sum total number of animals which are lactating and dry in a herd

Milk

It is a secretion of mammary glands in females. It is a source of nutrition to the young ones.

MOET

Multiple Ovulation Embryo Transfer is an composite technology which involves superovualtion, estrus synchronization among recipients, A.I. for donor, embryo recovery from donor and finally embryo transfer in the recipients females. MOET augments the genetic gain via utilising the two female paths (cow to cows and cows to bulls). The elite donors are hormonally treated to superovulate and A.I. is done and latter the embryos are flushed out and collected. The embryos are then transferred to surrogate cows of low genetic merit. This technique has advantage that it considerably increases the number of progeny obtained from the elite cows during her life time.

Mosaic

An individual or tissue having cells of two or more different genotypes.

Mosaicism

When an individual is composed of cells of more than one genetic background it is called a mosaic. Transgene integration takes place after the first round of cell division; thus, only a portion of the cells in the resulting individual will be transgenic. It is important that the germ cells carry the transgene so that subsequent offspring will also be transgenic.

Multivariate Analysis

Comprises a set of techniques dedicated to the analysis of data sets with more than one variable.

Mutation

Mutation describes the any change in the sequence of DNA.

National Research Council

NRC a scientific body in the U.S. that has regularly published nutrient requirements for animals based on the latest available research.

NDRI

National Dairy Research Institute, located in Karnal (Haryana), India.

Net Energy (NE)

The amount of feed energy actually available for animal maintenance and production, representing the energy fraction in a feed left after fecal, urinary, gas and heat losses are deducted from the gross energy value of a feed. Net energy can be further partitioned into the net energy necessary for maintenance, growth and lactation.

Network Project on Buffaloes

During the 8th five year plan (1993) a collaborated effort of various institutional and organised herds of buffalo initiated this programme. The CIRB, Hisar; PAU, Ludhiana; NDRI, Karnal; HAU, Hisar; IVRI, Izatnagar; and CBBF, Alamadi (TN); were collaborators for Murrah buffaloes while RAU, Vallabhnagar; GAU, Navsari and Anand; CCBF, Dhamrod and UAS, Dharwad were the collaborators for surti buffaloes. The aim was the sire evaluation and production of the elite germplasm.

Neutral Detergent Fibre

NDF the insoluble fraction containing all plant cell wall components left after boiling a feed sample in a neutral detergent solution. NDF is of low digestibility but can be broken down somewhat by the digestive tract micro organisms. NDF value is used to predict ruminant feed intake.

Non Protein Nitrogen

NPN nitrogen not derived from true protein, but usable by rumen microbes to build microbial protein, (e.g. urea).

NPCBB

National Project on Cattle and Buffalo Breeding was started in the year 2000 by Govt. of India for a period of 10 years, in two phases of five years each. Rupees 402 and 775.87 crore were allotted for the first and second phase respectively. The objectives of the NPCBB were (a) To arrange delivery of vastly improved artificial insemination service at the farmers doorstep. (b) To progressively bring under organized breeding through artificial insemination or natural service by high quality bulls, all breedable females among cattle and buffalo within a period of 10 years. (c) To undertake breed improvement programme for indigenous cattle and buffalo breeds so as to improve their genetic qualities.

Nucleus Breeding Scheme

The scheme can be closed or open depending upon the direction of germ flow.

Nucleus Herd

In a nucleus breeding scheme a central herd is created which is used entirely for the production of the males for breeding purpose in the population from the elite females.

ONBS

Open Nucleus Breeding System involves the two way transfer of genes form the nucleus herd to the farmers herd and also from the farmers herd to the nucleus herd, by introducing the superior animals form the other herd. This reduces the rate of inbreeding in the nucleus herd and increases the genetic progress because the superior animals are also available with farmers are used back in the nucleus herd.

Ova

Ova the female gamete, produced as a result of oogenesis from ovary.

Para Sexuality

It is a method of genetic recombination by means other than sexual and asexual reproduction. It was observed in certain microbes and recent form of transgenesis is also para sexual way of genetic transformation. It has opened avenues of gene mapping, revelation of fine gene structure, genetic dissection of diseases. Besides, principles of genetic mechanisms, cell growth, and differentiation, and mutagenesis, normal & abnormal human biology have become more amenable.

Peak Yield

Is the maximum yield or amount of milk produced in any day during lactation.

Persistency of Lactation

It is defined as 'the degree with which the rate of milk secretion is maintained during decline phase of the lactation with the advancement of lactation'. Mathematically, it is expressed as the average percentage or degree of decrease in milk production each month to that of the previous month.

Phenocopy

An individual whose phenotype has been modified by the environment by such an extent that it is similar to that produced by another, usually a mutant allele i.e. some times some environmental changes produce a trait /phenotype that resemble the phenotype produced by a specific gene e.g. Phocomelia in humans and Rumpless ness in poultry, caused by shaking the eggs at fetal development or injecting insulin in the developing human embryo.

Phenotype

The observable characteristics or appearance of an individual is called its phenotype. It is a sum total of genotype, environment and their interactions.

Physical Map

A map of the locations of identifiable landmarks on DNA (e.g., restriction enzyme cutting sites, genes), regardless of inheritance. Distance is measured in base pairs. For the human genome, the lowest-resolution physical map is the banding patterns on the 24 different chromosomes; the highest- resolution map would be the complete nucleotide sequence of the chromosomes.

Polymorphism

It refers to the simultaneous occurrence of more than two forms/ alleles in the population at the same time, such that rarest of rare can not be maintained by recurrent mutations. The frequency of Polymorphic allele is more than 0.01 and less than 0.99

Population

In genetics means the group of individuals (male and female animals) which can be breed among themselves.

Preservation

It means continued maintenance of a particular breed which is under the threat to extinct or has reached to an endangered level of population.

Preservative

A substance added to protect, prevent or retard decay, discolouration or spoilage of a substance during storage or use.

Progeny Testing

The criteria for selecting and evaluating the individual animal based on the performance of his progenies is known as Progeny testing.

Protein

A large molecule composed of one or more chains of amino acids in a specific order; the order is determined by the base sequence of nucleotides in the gene coding for the protein. Proteins are required for the structure, function, and regulation of the body cells, tissues, and organs, and each protein has unique functions. Examples are hormones, enzymes, and antibodies

QTL

A gene or a region of the chromosome affecting the phenotypic variation in continuously varying traits like milk production, disease traits and body conformation traits.

Qualitative Trait

A trait which shows a discontinuous variation and differ more in kind than in degree e.g. disease traits.

Quantitative Traits

Those traits which differ more in degree rather than in kind and show the continuous variation.

Replacement Rate

It is the percentage of female calves reached to AFC out of total female calve born or total calves born. It is an important consideration in a breeding system (progeny testing programme) to gain the better / higher response.

Response to Selection

Change in performance of progeny generation due to artificial selection is known as response to selection or genetic change or genetic gain. ($R = h^2 . S$ where h^2 is the heritability of the trait, and S is the selection differential)

Roughage

Feed high in fibre (greater than 18% crude fibre); tends to be bulky, coarse and low in energy.

Selection

Differential reproductive and survival rate i.e. it is a process of selecting the superior / best animals from the herd, whatever best may be, which will act as parents for the next generation in a herd.

Selection Accuracy

Is the correlation between the index value and additive genetic variance. It can also be obtained as a square root of heritability (h). Mathematically, $r_{IA} = h$

Selection Differential

The average phenotypic superiority of the selected parents over the population means is called as the selection differential ($S = M_s - P$ where, Ms is the mean of selected parents and P is the population mean before the selection)

Selection Index (I)

It is the best method of selection for animals where in the appropriate weights (b) are given to different traits and an index value is obtained by linear additive relationship among different traits i.e. the product of phenotypic value (x) and their weights are linear and additive in nature. $I = b_1x_1 + b_2x_2 + \ldots\ldots\ldots + b_nx_n$

Selection Intensity

It is the mean deviation of the selected animals in units of 6_p of the trait. It is also defined as the standardised selection differential. It depends on the proportion of animal selected, phenotypic standard deviation of the trait, number of traits under selection and their accurate measurements.

Selective Breeding

Is a breeding strategy for indigenous well defined breeds of cattle and buffalo through progeny testing programme or ONBS. Selective breeding strategy yield better response in field conditions.

Service Period or Days Open

It is the period from calving to conception.

Sire Evaluation

The result of progeny testing of bulls is reported in the form of an index, which indicates the genetic worth of an animal and is known as sire index. Based on sire index a numerical value is obtained indicating milk producing ability of a particular sire. The numbers of sires tested under progeny testing are evaluated based on sire index; this process is known as sire evaluation.

SNP

Single Nucleotide Polymorphism, pronounced as *snips.*

Somatic Cells

These are the cells of an organism other than those of the germlines.

Somatic Nuclear Transfer

In its most common form, this involves removal of the nucleus from a fertilized egg and replacement with the nucleus of a differentiated somatic cell such as a fibroblast. The result is a cloned embryo that has the genetic composition of the nuclear donor cell.

Species

In a classification of kingdom is further sub classified / divided into species and species are further divided into breeds. There is a discontinuity of reproduction between two species, which means the two species do not interbred or the progeny of two species do not produce the fertile offspring when they are further mated.

Sperm

It is the male gamete, produced as a result of spermeogenesis from testis.

Spermatogonia

Early developmental precursors of sperm that can be removed from the testes, transfected with a transgene, and then re-implanted to develop into mature sperm that carry the transgene.

Standard Deviation

It is an important measure of dispersion and is defined as the root mean squared deviation of the sample.

Stem Cells

The progenitor cells that are undifferentiated cell which upon a proper stimulus has a capability to differentiate into any type of cell. They can renew themselves indefinitely. Those originated from the inner cell mass of blastocyst are known as embryonic stem cells.

Superovulation

Is a term used to describe the drug induces production of multiple eggs for use during assisted reproduction. The superovualtion is used to produce multiple embryos.

Supplement

Feed or feed mixtures used to improve the nutritional value of basal feeds. A supplement is rich in one or more of protein, energy, vitamins, minerals or antibiotics, and is combined with other feeds to produce a more complete feed. Often used interchangeably with concentrate

Sustainable Development

It is the management and conservation of the natural resource base and orientation of the technological and institutional change in such a way as to ensure the attainment and continued satisfaction of human/animals needs for present and future generations

Synchronization

Synchronization of estrus refers to the process of scheduling a group of embryo recipient females to come into estrus at approximately the same time as the superovulated embryo donor animal.

Tandem Repeat Sequences

Multiple copies of the same base sequence on a chromosome; used as a marker in physical mapping

Total Digestible Nutrients (TDN)

TDN is the energy value of feedstuffs, comparable to digestible energy in accuracy. TDN over-estimates the energy value of roughages in comparison to grains

Transgenesis

It is a technique which permits manipulation of gene from one organism to be introduced into the genome of other organisms.

Transgenic Animals

Those animals which receives transgene or whose genetic composition has been altered by addition of foreign DNA; they are also defined as animals containing recombinant DNA in their genome that were introduced by human intervention. Transgenic animals express protein or other trait from animal of same species or different species as a result of stable incorporation of one or more foreign genes into its genome.

True Breeding Organisms

True breeding organisms are those which are homozygous for the trait under consideration.

Variance

It is an important measure of dispersion. It is used to measure the variation among the animals and is defined as the mean squared deviation of the sample. It can also be obtained by squaring the standard deviation of the sample.

Vulnerable Species

Those species or breeds are considered to be vulnerable if their population goes less than 5,000 animals.

Wallowing Behaviour

It is characteristic of Indian buffaloes. Wallowing has two purposes; the most obvious is that of cooling, the other is protection from insects. Wallowing during daytime is done during the hottest hours. Wallowing during night time is instead a way for the animal to protect itself from insects.

Wet Average

It is defined as the total milk given by lactating animals, divided by their number in the herd. It is the average of animals that are in milk and does not include the dry animals.

Zygote

Zygote is produced by fusion of two gametes (fusion of sperm with egg).

❑❑❑

References

Abha Chhabra, K. R. Manjunath, Sushma Panigrahy and J. S. Parihar (2009). Spatial pattern of methane emissions from Indian livestock. *Current Science*, vol. 96, no. 5, 10 March 2009: 683:689

Abo-Omar, J. M. 1989. Methane losses by steers fed ionophores singly or alternatively. W.D. Dissertation. Colorado State Univ., Fort Collins.

Alderson, L. (1981). *FAO Animal Production and Health Paper*, 24: 53-76. Rome, FAO.

Anderson, K L., T. G. Nagaraja, J. L. Morrill, T. B. Avery, S. J. Galitzer, and J. E. Boyer. 1987. Ruminal microbial development in conventionally or early-weaned calves. *J. h i m . Sci.* 64:1215.

Anderson, T. W. (1958) *An Introduction to Multivariate Statistical Analysis.* Wiley, New York.

Anonymous (2003). Seventeenth livestock census, *Govt. of India*, New Delhi, India.

Anonymous (2007). Dairy India Yearbook. *Sixth ed.*, Published by A-25, Priyadarshini Vihar, New Delhi, India.

Balain, D.S. (1988). Progeny testing, its present status and future options for buffalo development. *Indian J. Anim. Prod. Mgmt.* 4 (3): 334-342.

Balakrishnan, M and Nagarcenkar, R. (1988) Effect of cooling by spraying water, wallowing and mud plaster on buffaloes during summer. *Pro. of second world buffalo congress* held at New Delhi, 1988 pp: 317-322

Balaraman, N. (2006) A buffer to sustain livelihood. *The Hindu survey of Indian Agriculture* 2006, pp: 135

Bartlett, M.S. (1937). Some examples of statistical methods in research in agriculture and applied biology. *J. Roy Stat. Soc Supp.*, 4, 137-183

Basu, S.B and Mahla, J.S (1980). Genetic and financial consequences of progeny testing in buffaloes. *Indian J. Anim. Sci.* 50(10):805-809

Beever, D. E., S. B. Camrnell, J. D. Sutton, M. C. Spooner, M. J. Haines, and J. I. Harland. 1989. The effect of concentrate type on energy utilization in lactating cows. *Proc. 11th Symp. on Energy Metabolism EAAP* Publication No. 43:33.

Bhat P.N., Taneja V.K. and Garg R.C. Effect of crossbreeding on reproduction and production traits. *India J. Anim. Sci.* 48: 71-78.

Blaxter, K. L. 1989. Energy Metabolism in Animals and Man. *Cambridge University Press,* New York.

Blaxter, K. L., and J. L. Clapperton. 1965. Prediction of the amount of methane produced by ruminants. *Br. J. Nutr.* 19:511.

Cammell, S. B., D. E. Beever, K. V. Skelton, and M. C. Spooner. 1980. The construction of open circuit chambers for measuring gaseous exchange and heat production in sheep and young cattle. *Lab. Pract.* 30:115.

Carmean, B. R. 1991. Persistence of monensin effects on nutrient flux in steers. M.S. Thesis. *Colorado State Univ.,* Fort Collins.

Chacko, C.T, Walter Diener and T.C. Mathew. (1984). Sire evaluation under field conditions in Kerala: The influence of environmental effects on lactation. *Indian Vet. J.* 61: 1034-1042.

Chatfield, C. and Collins, A.J. (1990). *Introduction to Multivariate analysis.* Chapman and Hall publications

Chauhan, V.P.S, Godbole, A.D., Saigal, A.K. Gunjal, M.G and R.G. Kulkarni. (1994). *Indian J. Dairy Sci.*, 47 (7): 534-535.

Cochran, W.G. (1934) The distribution of quadratic forms in a normal system with applications to the analysis of variance. *Proc. Cambridge Philos. Soc.,* 30, 178-181

Czerkawski, J. W. 1986. An Introduction to Rumen Studies. Pergamon Press, New York.

Czerkawski, J. W., and G. Breckenridge. 1977. Design and development of a long-term rumen simulation technique (Rusitec). *Br. J. Nutr.* 38:371.

Czerkawski, J. W., K. L. Blaxter, and F. W. Wainman. 1966. The metabolism of oleic, linoleic, and linolenic acids by sheep with reference to their effects on methane production. *Br. J. Nutr.* 20:349.

Deb, S.M., M. Gurnani and C.T. Chacko. (1998). Comparision of sire evaluation methods under field condition. *Indian J. Dairy Sci.*, 51(4): 210-215.

Dhara, S.K. and Chakravarty, A.K. (1996) Principal Component Analysis- A new approach for prediction of breeding value of milk production in Murrah buffaloes. *In proceeding second national symposium on buffalo research for higher productivity*, CCSHAU, Hissar, 5-6 April, pp 6

Dorf, R.C. *Matrix Algebra – A programmed Introduction.* Wiley, New York.

EPA. 1993. Anthropogenic methane emissions in the United States: Estimates for 1990. Report to Congress. EPA 430-R-93-003. Flatt, W. P,, P. J. Van Soest, J. F. Sykes, and L. A. Moore. 1958. Adescription of the energy metabolism laboratory at the US. Department of Agriculture, Agricultural Research Center inBeltsville Maryland. In: G. Torbek and H. Aersoe (Ed.) 1. Symposium on energy metabolism: Principles, methods and general aspects. Statens Husdyrbrugsudvalg, Copenhagen, Denmark. 53.

Falconer, D. S. and Mackay, T.F.C.,(1994) Introduction to Quantitative Genetics, 4th *edition*, Longman Publication.

FAO (2002) www. FAO. org/ dad/ -is/

Gaur, G.K. (2001). Genetic consequences of field progeny testing scheme in crossbred cattle. *Indian J. Dairy Sci.*, 54 : 47-49.

Godbole, A.D, Chauhan, V.P.S, Saigal,A.K. Gunjal, M.G and C.R. Gujarathi. (1994). Partitioning of environmental variation in field progeny testing data. *Indian J. Anim. Sci.*, 64(10): 1067-69.

Goodrich, R. D., J. E. Garrett, D. R. Gast, M. A. & rick, D. A. Larson, and J. C. Mieske. 1984. Influence of monensin on the performance of cattle. *J. Anim. Sci.* 58:1484.

Haaland, G. L. 1978. Protected fat in bovine rations. Ph.D. Dissertation. Colorado State University, Fort Collins.

Handique, N. (1976). M.Sc, Thesis submitted to National Dairy Research Institute, Karnal

Henderson, C.R. *Linear Models in Animal Breeding* Wiley, New York.

Hinks, C.J.M (1970). The selection of dairy bulls for artificial insemination. *Anim. Prod.* 12: 569

Hutcheson, J. P. 1994. Anabolic implant effects on body composition, visceral organ mass and energeticosf beef steers. Ph.D.Dissertation. Colorado State Univ., Fort Collins.

ICMR recommendations, (1989) Sukumar, De. (2002). Cited in Outlines of Dairy Technology *17th edition*, Oxford University press, New Delhi.

Itabashi, H., T. Kobayashi, and M. Matsumoto. 1984. The effects of rumen ciliate protozoa on energy metabolism in some constituents in rumen fluid and blood plasma of goats. *Jpn. J. Zoo Tech. Sci.* 55:248.

Jobsons, S.D. *Applied Multivariate Data Analysis Statistics.* (2nd Volume) Wiley, New York

Johnson, D. E., T. M. Hill, G. M. Ward, K. A. Johnson, M. E. Branine, B. R. Carmean, and D. W. Lodman. 1993b 13. Principle factors varying methane emissions from ruminants and other animals. In: M.A.K. Khalil (Ed.). Atmospheric Methane: Sources, Sinks, and Role in Global Change. NATO A D 1 Series Vol 113, Springer-Verlag, Berlin, Germany.

Johnson, R.A. and Wichern, D.W. (1996). Applied multivariate statistical analysis. *Prentice-Hall of India Private Limited.*

Kaushik, S.N. and R.C. Garg. (2000). Present status of field progeny testing programme of PDC Meerut. *Indian Dairyman*, 52 (8): 13-16.

Kujawa, M. A. 1994. Energy partitioning in steers fed cottonseed hulls and beet pulp. Ph.D. Dissertation. Colorado State Univ., Fort Collins.

Lush, J. L. (1945). *Animal Breeding Plans, 3rd edition,* Iowa State University Press, Ames, Iowa (USA).

Maijala, K., Chevekaev, A.V., Devillard, J.-M., Reklewski, Z., Rognoni, G., Simon, D.L. & Steane, D.E. (1984). Conservation of Animal Genetic Resources in Europe (EAAP working party report). *Livestock Prod. Sci.*, 11: 3-22.

Mallows, C.L. (1973). Some components on Cp *Technometrics*, **15** : 661-675.

Mehla, J.S., Basu, S.B and Raina, V.S.(1981) The cost of testing buffalo bulls. *Indian.J.Anim.Sci.*51:135

MW, P. W., and H. F. Tyrrell. 1979. Methane production in dairy cows. J. Dairy Sci. 62:1583.

Nagarcenkar, R. and M.K. Rao. (1983). Progeny testing in the field conditions, *Indian Dairyman*. 35: 261-266.

Nagarcenkar, R., Bhatnagar, D.S., Gurnani, M. and Rao, M.K. (1981). Crossbreeding of Tharparkar with exotic breeds. *Annual Report 1981*, NDRI, Karnal, p. 68.

NBAGR, Karnal (2004). National Training Programme on sustainable production in farm animals through breeding and management intervention. Compendium of Lectures.

Okine, E. K., G. W. Mathison, and R. T. Hardin. 1989. Effects of changes in frequency of reticular contractions on fluid and particulate passage rates in cattle. *J. h i m . Sci.* 67:3388.

Oltenacu, P. A., and C. W. Young. (1974). Genetic optimization of a young bull sampiing program in dairy cattle. *J. Dairy Sci.* 57: 894

Park, H. S., J. J. Ramsey, J. Torrent, and D. E. Johnson. 1994. Nutritional evaluation of an experimental fat supplement for cattle. Metabolic Laboratory Report, Colorado State University, Fort Collins.

Pindyck, R.S. and Rubinfeld, D.L. (1981). Econometric Models and Economic Forecast. *2nd edition*. McGraw Hill International Book Co. Singapore.

Rumpler, W. V., D. E. Johnson, and D. B. Bates. 1986. The effect of high dietary cation concentration on methanogenesis by steers fed diets with and without ionophores. *J. Anim. Sci.* 62:1737.

Saa, C. F., T. M. Hill, and D. E. Johnson. 1993. Persistence of methane suppression by ionophore and glycopeptide in steers fed a brome hay diet. CSU Beef Program Report.

Sastry, N.S.R and Thomas, C.K (2005) *Livestock production Management*. Kalyani publisher New Delhi

Searle, S.R. (1971) *Linear Models* Wiley, New York

Sharma, R.C. (1985) Breeding plan for optimizing genetic gain in a zebu breed. Ph.D. Thesis submitted to Punjab University

Sinha, Soniya, Chakravarty, A.K. and Joshi B.K. (2000). Genetic distance analysis of Karan Fries cows in relation to level of heterzygosity and milk production. *Indian J. of Dairy Science.*

Snedecor, G.W. and Cochran, W.G. (1994). Statistical methods. 8th Ed. *Iowa State University Press, Ames,* USA.

Soller, M., Bar Anan-R and Pasternak, H (1966) Selection of dairy cattle for growth and milk production. *Anim. Prod.* 8:109-119

Stumm, C. K., H. J. Gijzen, and G. D. Vogels. 1982. Association of methanogenic bacteria with ovine ciliates. *Br. J. Nutr.* 47:95.

Sunderasan, D. (1975) *Livestock Breeding in India.* Vikas Pub. House, New Delhi.

Swift, R. W., J. W. Bratzler, W. H. James, A. D. Tillman, and D. C. Meek. 1948. The effect of dietary fat on utilization of the energy and protein of rations by sheep. *J . h i m . Sci.* 7:475.

Taneja V.K., Bhat P.N. and Garg R.C. Genetic divergence in various Sahiwal x Holstein crossbred grades. *Theor. Appl. Genet.* 54: 69-74.

Thomas, Preeti. R. and Chakravarty, A.K. (1999). Canonical correlation analysis for studying the association of breeding efficiency and expected producing ability with growth and reproductive traits of Murrah buffaloes. *Indian J. of Dairy Science.*

Tomar, S. S. (2004). *Text book of Animal Breeding*. Kalyani Publications, New Delhi, India.

Van der Honing, Y., B. J. Wieman, A, Steig, B. van Donselaar. 1981. The effect of fat supplementation of concentrates on digestion and utilization of energy by productive dairy cattle. *Neth. J. Agnc. Sci.* 29:79.

VanVleck, L.D. (1964) Sampling the young sire in artificial insemination. *J. Dairy Sci.* 47: 441

VanVleck, L.D. (1986) Evaluation of dairy cattle breeding programs: specialized milk production. *Proc. 3rd world congr. Genet. Appl. Livest. Prod.* 9, 141-152.

Wainman, F. W., P.J.S. Dewey, and A. C. Brewer. 1984. Feedstuffs evaluation unit. Fourth Report. Rowett Research Institute, Aberdeen, Scotland.

Weller, J. L., Soller, M. and Broody, T. (1988). Linkage analysis of quantitative traits in an interspecific cross of tomato (*L. esculentum x L. pimpinellifolium*) by means of genetic markers. *Genetics* 118, 329 – 339.

Whitelaw, F. G., W. J. Eadie, L. A. Bruce, and W. A. Shand. 1984. Methane formation in faunated and ciliate-free cattle and its relationship with rumen volatile fatty acid proportions. *Br. J. Nutr.* 52:261.

Wolin, M. J., and T. L. Miller. 1988. Microbe interactions in the rumen microbial ecosystem. In: P. N. Hobson (Ed.). The Rumen Ecosystem. Elsevier Applied Science, New York.

www.dahd.nic.in/ (Dept. of Animal Husbandry, GOI, New Delhi)

Yadav, R.S. and Gulhati, H.K (2006) Management and buffalo production system in India. Pro. of the National symposium on Buffalo for Rural upliftment. held at Mumbai from 27-30 May 2006 pp:146-158.

□□□

Index

A

B

C

D

E

F

N

O

P

Q

R

S

T

U

V

W

□□□

About the Authors

Dr. Vikas Vohra is working as Assistant Professor / Jr. Scientist in the AGB Division of F.V.Sc &AH, SKUAST – Jammu since, 2005. He has done his Graduation degree (B.V.Sc & A.H) in 2002 from the first Veterinary College of Independent India which is College of Veterinary and Animal Sciences, Mathura (U.P), which is at present known as UP Pandit Deen Dayal Upadhyay Pashu Chikitsa Vigyan Vishwavidyalay Evam Go Anusandhan Sansthan, Mathura. He obtained his Post Graduation degree in the discipline of Animal Genetics and Breeding from A.G. Division of Indian Veterinary Research Institute (IVRI), Izatnagar, Bareilly (U.P) in 2004. He was awarded his PhD degree in Animal Genetics and Breeding from Dairy Cattle Breeding Division of National Dairy Research Institute (NDRI), Karnal (Haryana) in the year 2010. Through out his Academic career he was a meritorious student and NDRI topper student during his PhD programme. He was awarded with ICAR fellowship, CSIR-NET LS, ICMR fellowship and Institute fellowships during his professional degree programmes. He has four International and various national research publications in the peer reviewed journals of repute, collaborated in various practical and training manuals, and guiding M.V.Sc students. He was an honoured member of Academic council of NDRI (2008-09). He is a privileged reviewer of British Poultry Science Journal. His research contribution includes identifying new molecular and breeding strategies to increase the milk production in cattle and buffaloes.

Professor (Dr.) A. K. Chakravarty, Fellow of National Academy of Dairy Science, is working as Principal Scientist and In-charge of Artificial Breeding and Research Center at National Dairy Research Institute, Karnal. He completed his Masters Degree in the field of Animal Genetics and Breeding from Haryana Agricultural University, Hisar in 1982. He joined the ICAR in Agricultural Research Services in 1985 as Scientist S-1 at Central Institute for Research on Buffalo, Hisar. He obtained his PhD degree in the field of Animal Genetics and Breeding from Haryana Agricultural University, Hisar in 1986. He joined the National Dairy Research Institute (NDRI), Karnal in 1988. He has wide experience in handling several institutional, inter-institutional and externally funded projects including the Co-Principal Investigator of National Agricultural Technology Project on 'Sustainable High Production of Indigenous and

Crossbred Cattle and Buffaloes in rural households under rice-wheat production system' funded by World Bank in Indian Council of Agriculture Research. He has guided many M.Sc / M.V.Sc and Ph.D students in Animal Genetics and Breeding at NDRI, Karnal. Prof. (Dr.) Chakravarty has more than 22 years of teaching experience in Animal Genetics and Breeding at NDRI, Karnal and was awarded national fellowship and acted as an expert for different institutions/organizations. He is engaged in the cattle and buffalo breed improvement programmes of several states of India. His research publications have been published in various refereed national and international journals.